Advance praise for *Resilient Agriculture*

As we start to change the weather, resilience will become a watchword for farmers, as this fine book demonstrates. It's strong advice—and it reinforces the essential truth, which is that we must keep climate from changing too much—because there's nothing even the best farmer can do to cope with a truly overheated planet.

— Bill McKibben, author, *Deep Economy*

Lengnick uses her wide-ranging scholarship to locate current food systems in time and space and asks farmers and ranchers who are creating new food systems that are more climate and community friendly to tell their stories of what they are doing, how they are doing it, and why. This book is accessible and compelling—a must read that builds hope for systemic change for a more sustainable future.

— Cornelia Butler Flora, Charles F Curtiss Professor Emeritus,
Sociology and Agriculture and Life Science, Iowa State University;
and Research Professor, Kansas State University

Industrial farmers must continue relying on the government. Sustainable farmers must learn to accommodate the vagaries of nature—including changes in climate. Stories of progressive farmers who have found ways of coping, which others eventually must learn, highlight this comprehensive review of agricultural resilience and sustainability.

— John Ikerd, Professor Emeritus of Agricultural
and Applied Economics, University of Missouri

In this timely and well-written book, Laura Lengnick combines the latest science with a search for solutions. She finds answers in the fields and pastures of some of the most innovative sustainable agriculturalists in the country. Our food future hinges on their experiential, local knowledge about how to manage for resilience. Without a doubt, I'll be using this important book in my teaching right away.

— Neva Hassanein, Professor of Environmental Studies,
University of Montana; and author, *Changing the Way America Farms:*
Knowledge and Community in the Sustainable Agriculture Movement

Laura Lengnick has researched our changing weather patterns well. With that data as a background you can read the stories of award winning farmers and learn how they have adapted to the changes they've experienced and their plans for the future. No doubt, this book will help you identify problems you might encounter and give you ideas to help you plan a resilient future for your farm.

— Cindy Conner, Homeplace Earth, author,
Grow a Sustainable Diet and *Seed Libraries*

Farmers now need to design a resilient, regenerative agriculture for long-term economic returns. Laura Lengnick's new book provides a comprehensive analysis on how to begin that journey. A must read for anyone interested in the future of farming.

— Frederick Kirschenmann, author,
Cultivating an Ecological Conscience: Essays From a Farmer Philosopher

This is a timely and important contribution to bringing the science and thinking behind resilience to address the practical problems facing agriculture under global change. A focus on production and efficiency alone reduces capacity to deal with the rising frequency of unexpected secondary effects and novel situations. Laura Lengnick's book brings the two needs together in a readable and helpful way.

— Brian Walker, Resilience Alliance Fellow

Readers looking for a ray of hope buried in the avalanche of bad news from climate science will not be disappointed. Dr. Lengnick has uncovered a collection of "new pioneers" who are not backing down from daunting challenges to their vision of a better way of producing food in a changing climate.

— Eugene S. Takle, Professor of Atmospheric Science and
Pioneer Hi-Bred Professor of Agronomy, Iowa State University

Anyone interested in the interface between our food supply and climate change whether at the consumer, practitioner, activist or political level will greatly benefit from the insights in this important work.

— Marc N. Williams, Ethnobotanist,
Executive Director, Plants and Healers International

Resilient Agriculture is an important and timely book, meshing concept and practice. The description of climate change is nuanced and complete, but more importantly it provides the structure for farms and ranches to be adaptive and resilient. The second half of the book, profiling innovation across products, geography, and scale, is unique. The focus on solutions rather than problems, building on the experience of farmers and ranchers, is exactly what is needed.

— Tim Griffin, Director, Agriculture Food and Environment,
Friedman School of Nutrition Science and Policy, Tufts University

This timely and important book synthesizes vital wisdom that our society needs to weather the shocks of climate change. It brings the experiences of some of the best farmers, growers and ranchers in the country, as they cope with changes that they are already observing on their farms and ranches.

— Dr. Molly D. Anderson, Partridge Chair in Food & Sustainable
Agriculture Systems, College of the Atlantic, Bar Harbor, Maine

RESILIENT AGRICULTURE

CULTIVATING FOOD SYSTEMS FOR A CHANGING CLIMATE

LAURA LENGNICK

Cover design by Diane McIntosh.
All images © iStock (Orchard: Rike_; Maize field: Belinda Pretorius;
Farm and tractor: Richard Goerg; Cows: Dutch Scenery)

Printed in Canada. First printing April 2015.

New Society Publishers acknowledges the financial support of the Government of Canada through the Canada Book Fund (CBF) for our publishing activities.

Any other inquiries can be directed by mail to:

New Society Publishers
P.O. Box 189, Gabriola Island, BC V0R 1X0, Canada
(250) 247-9737

LIBRARY AND ARCHIVES CANADA CATALOGUING IN PUBLICATION

Lengnick, Laura, author
Resilient agriculture : cultivating food systems for a changing climate / Laura Lengnick.

Includes index.
Issued in print and electronic formats.
ISBN 978-0-86571-774-9 (pbk.).— ISBN 978-1-55092-578-4 (ebook)

1. Sustainable agriculture. 2. Agriculture—Environmental aspects.
3. Climate change mitigation. 4. Climatic changes. I. Title.

S494.5.S86L45 2015 630 C2015-902027-1
C2015-902028-X

New Society Publishers' mission is to publish books that contribute in fundamental ways to building an ecologically sustainable and just society, and to do so with the least possible impact on the environment, in a manner that models this vision. We are committed to doing this not just through education, but through action. The interior pages of our bound books are printed on Forest Stewardship Council®-registered acid-free paper that is 100% post-consumer recycled (100% old growth forest-free), processed chlorine-free, and printed with vegetable-based, low-VOC inks, with covers produced using FSC®-registered stock. New Society also works to reduce its carbon footprint, and purchases carbon offsets based on an annual audit to ensure a carbon neutral footprint. For further information, or to browse our full list of books and purchase securely, visit our website at: www.newsociety.com

Contents

Acknowledgments

This book emerged as a result of coincidence, design, curiosity, respect, patience and a lot of support. It was coincidence that introduced me to resilience thinking and the Transition Initiative in the same year. By design, the third National Climate Assessment empowered a broad diversity of voices to engage in a national discussion on the realities of climate change. Curiosity, my own and that of many others, enriched my exploration of how sustainability and resilience are the same, how they are different and how we might use each to sustain our communities into the 21st century. My abiding respect for sustainable farmers, growers and ranchers made me certain that they had something unique to contribute to the national discussion on agricultural adaptation. I had to be patient with myself and ask for the patience of many others as this project took on a life of its own and just wouldn't be finished until it was done. And through it all, I enjoyed the support and encouragement of family, friends and a lot of wonderful souls well-met during the course of this work.

Friends and colleagues reviewed drafts and made excellent suggestions for improvement, including Mara Shea, Katy Estrada, Bob Sigmon, Kate Wheeler, Mallory McDuff, Megan Cornett Leiss, Lyle Estill, Tami Schwerin, Alison Perrett, Charlie Jackson, Jay Bost, Jim Worstell, Anna Nassiff, Lisa Johnson, Kim Kroll, Amy Boyd and Mark Siler. Dennis Merritt, Preston Sullivan, Elizabeth Henderson, Alisa Hove, Patrick Sweatt and Michelle Miller were especially generous with their time and thoughtful critique. Mallory McDuff and Catherine Reid were a constant inspiration and Karen Gaughan never missed a beat. Charlie Walthall, Jerry Hatfield, Peter Backlund and Elizabeth Marshall taught me much in our year of rich collaboration. The staff at New Society supported me with grace and good cheer throughout the sometimes bumpy learning process that is a first book. Filo and the West End Bakery offered exactly the right vibe for productive writing. Many others assisted along the way with ideas, introductions and information that helped me make sense of what I was learning as I worked through this project. I thank you all.

Special thanks to the good folks at Abundance Foundation, especially Lyle Estill and Tami Schwerin, who opened the way for this book and convinced me that I was the person to write it. My deep appreciation goes out to the farmers, growers and ranchers who so freely gave of their time and were willing to talk about their experiences of farming in a changing climate at a time when it was still a bit risky to do so.

And finally, I want to thank my family for giving me the space I needed to write this book, for understanding when I missed family get-togethers and for their patience as I went deeper and deeper into this work. To my husband Weogo, special thanks and abiding gratitude for everything, but especially for never, not even once, rolling your eyes when I asked if something could wait until "after the book is done."

To Kade Carson, Max Lengnick,
and the rest of their generation.
May your parents and grandparents
find the love, wisdom and courage
to put your world on a sustainable
and resilient path.

CHAPTER 1

Sustaining Agriculture in a Changing Climate

Climate change is upon us, and agriculture is inextricably involved. Fundamental to our identity as a species, crucial to the health and well-being of our communities, the way that we eat fuels the 21st-century challenges that threaten our way of life. How can we resolve this dilemma? The path that we have taken as a species shapes the possibilities of our future.

For the last ten thousand years, we have been on a long walk out of the Earth's wild places, onto the farm and into the city. In the slow evolution from foraging to agriculture, humanity took up, quite unconsciously, the work of nature. We took on the work of cultivating plants and animals, of caring for them and guiding their evolution. With this change in the way we related to the ecosystems that sustained us, we changed ourselves, and we changed the Earth.[1]

Agriculture emerged as our dominant food acquisition strategy rather late in our history as a species. For millions of years, bands of *Homo sapiens* adapted to the available food resources produced by the native ecosystems in which they lived. Satisfying this basic need to eat drove the evolution of a wide diversity of foraging strategies shaped largely by local ecological conditions.

Archeological evidence suggests that by the end of the last ice age, about twenty thousand years ago, humans had evolved a comfortable way of life based on foraging for food. In diverse ecosystems across the Earth, humans, like all animals, were part of the native food web. Food foraging sustained stable human populations for millennia, but about ten thousand years ago, something changed. All over the Earth, human population began to grow. Wherever that happened, you find agriculture.

Early Systems of Agriculture

Agriculture can be defined as the cultivation of deliberately bred crops and livestock for food, fiber and other materials. It evolved as a distinctly human form of food acquisition during the Neolithic Period (10,000–2000 BCE), as foraging cultures slowly came to rely on one of three distinct forms of early agriculture.

In the desert grasslands of the world, ecosystems too dry for the cultivation of plant foods, humans first hunted and then domesticated grazing animals—sheep, goats, cattle and camels—to sustain themselves. This led to the evolution of *pastoralism* as an efficient means of food acquisition. In the wetter forest ecosystems, humans first only gathered wild foods, but eventually began to care for some of the plants they favored. These *horticulturalists* made use of the diversity of edible plants and small forest-dwelling animals—swine, poultry and guinea pigs—and managed gardens of favored plants and animals in clearings rotated through their forested homelands. In just a few unique places, where the right combination of natural resources came together—plants, animals, land, water—a third form of early agriculture evolved. *Sedentary agriculture* emerged in the great river valleys of savanna ecosystems inhabited by a diversity of plants that produced edible fruits, seeds and vegetables, as well as large herds of grazing animals. Sedentary agriculturalists created the first permanent farms producing domesticated grains and livestock on land kept fertile by yearly flooding.

Even though these three forms of agriculture evolved in very different ecological circumstances, they share a number of characteristics. First, all are embedded in local ecosystems and dependent to a great degree on ecological processes to provide water, nutrients, pest suppression,

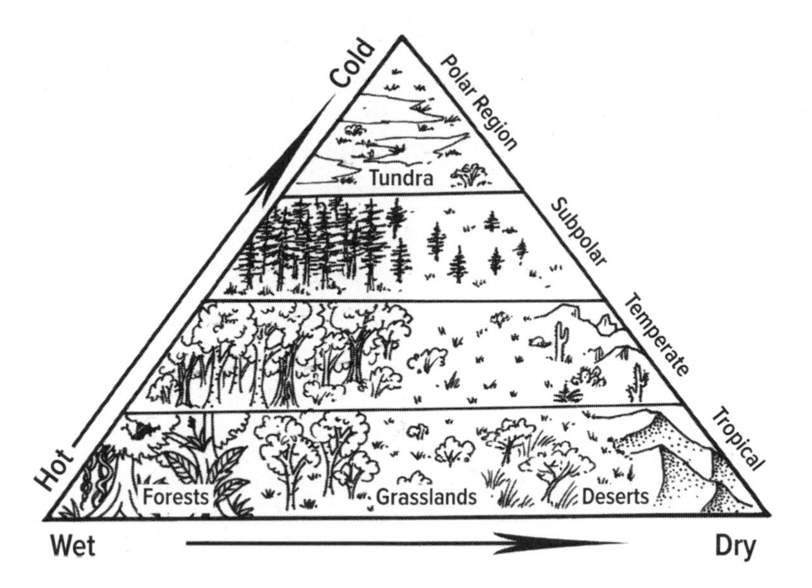

FIGURE 1.1. Major Terrestrial Biomes. The land-based ecosystems of the earth can be grouped into four major biomes—tundra, forest, grasslands and desert—based on dominant vegetation and climate. Pastoralism evolved in grasslands, shifting horticulture evolved in forests, and sedentary agriculture evolved in both temperate and desert regions with access to manageable water supplies. Credit: Sherri Amsel

waste disposal and other services needed to successfully produce edible plants and animals. Early farmers increased food production in these systems in two ways. They could care for favored plants and animals by improving growing conditions, providing food and reducing the risk of predation. And they could improve the yield and quality of the food produced by the species under their care through careful selection for desired qualities and yield. But in every essential way, these early systems of agriculture were adapted to ecological resource limits.

All these early systems of agriculture took advantage of the many benefits that animals provide to food production. Most importantly, animals are an efficient strategy for producing food from plants that are inedible to humans, for gathering and storing foods, and for utilizing food wastes. The ruminant animals—sheep, goats, cattle, camels—convert inedible grasses and forbs into high-quality meat, milk and eggs for human consumption, while others—dogs, swine, poultry—can consume a wide variety of foods and make great garbage collectors. Animals are excellent "biological silos" for storing excess production. Ruminants are particularly useful as buffers in extreme conditions because they can survive for long periods without food. At this point in the development of agriculture, animals were not yet used for traction.[2]

All of the early agricultures created an energy profit—more food energy was produced than energy invested in production—although none came close to the energy profit estimated for foraging of forty food calories produced for every calorie of labor invested.[3] The energy profit in pastoral and horticultural systems has been estimated at about eleven food calories harvested for every calorie of labor invested. Because sedentary agriculture required more intensive management, it is the least profitable of the early agricultures, reaping about eight calories for every labor calorie invested.

Foraging and the early farming systems that followed it propose very different solutions to the same basic question facing all animals: How best to allocate the available time and resources to acquire food? All things being equal, animals (including humans) tend to solve this effort-allocation problem by maximizing the capture of calories, protein and other desired foods in a way that yields the most return with the greatest certainty in the least time for the least effort.[4] Moderate, reliable returns are usually preferred over fluctuating high returns. It turns out that, for a long time, foraging was a good solution to the effort-allocation problem facing early humans. But climate change changed everything.

At the end of the last ice age, between ten and twelve thousand years ago, agriculture began to replace foraging as a food acquisition strategy in many regions. Rapid warming and increasing climatic variability caused ecological disturbances that changed the effort-allocation equation. Climate change led to shifts in the ranges of plants and animals, changing the mix of available food species and causing plants and animals that could not adjust to the new climate conditions to disappear.

At the same time, lowland and coastal areas were being flooded by rising seas fed by the melting of massive continental glaciers. Archeological evidence suggests that during this time foraging cultures throughout the world began to broaden the number of plants and animals they used for food and started to cultivate favored plants and animals. The first evidence of warfare dates from these challenging times as well. People across the globe shifted from food gathering to cultivating crops and livestock during this time, though some foraging cultures persist to this day.[5]

Ultimately, the effort-allocation problem boils down to a question of managing five basic natural resources: land, water, labor, plants and animals. Pastorialists focused on managing livestock because grasslands have few native edible plants and shifting, uncertain rainfall; however, grasslands have a lot of grass and mobile grazing animals whose milk, blood and meat provide sustenance to humans. Horticulturalists did not have benefit of large domesticable animals, but had large expanses of forest filled with edible plant foods and plentiful rainfall, and they focused on cultivating native edible plants and small woodlands animals. Sedentary agriculturalists had benefit of a diversity of edible grasses and other edible plants as well as domesticable animals, and they developed agricultural systems that used both. All three forms of early agriculture increased food supplies, causing human populations to rise, but it was sedentary agriculture that ignited the human population explosion that continues to this day.

The first clear evidence of fully developed systems of sedentary agriculture emerged in several locations in Asia about six thousand years ago. The sedentary agriculturalists inhabiting the Fertile Crescent were originally pastoral cultures that kept livestock and foraged wild foods, including native grasses. These native grasses flourished under the changing climate conditions, offering pastoralists a new food source and encouraging innovations in the tools and technologies needed to cultivate wild grains as a dietary staple. The people of the Fertile Crescent domesticated the abundant native grasses and developed the first grain crops—early forms of wheat and barley.

By 3000 BCE, the sedentary agriculturalists had increased agricultural productivity by using natural resources—particularly land and water—more intensively. The development of irrigation works supported huge increases in yield and stabilized production. Animal power provided labor, and animal manures were used as fertilizers to replace nutrients taken from the soil in harvested crops. This intensification of sedentary agriculture produced large surpluses of food but cut the energy profit of agriculture in half—from eight to four food calories for every labor calorie invested[6]—because of the additional work required to build, maintain and power the production system with human and animal labor. Intensive sedentary agriculture was so productive that fewer people were required for food acquisition, freeing up some to devote all of their time to other pursuits. This release of human resources from food production sparked the development of the first cities and gave birth to human civilization as we know it. All because of climate change.

American Agriculture

Up until about a hundred and fifty years ago, sedentary agriculture really had not changed all that much since Neolithic times. Neolithic farmers domesticated just about every animal and plant that could be domesticated. These early agriculturalists developed irrigation, used animal manures as fertilizers, rotated grain and legume crops and used pesticides. Over the last five thousand years, in every favorable biome, regardless of culture, the basic elements of intensive sedentary agriculture remained the same.

Contemporary American farmers and gardeners will recognize the practices used by Roman farmers in about 200 CE. To enhance soil fertility, the Romans cultivated grain crops like spelt, barley and wheat in rotation with alfalfa and mixed cover crops of beans, vetch, chickpea and clover. They managed sheep, goats, pigs and cattle on pasture and gathered hay from wild grasslands and acorns from oak forests to feed their livestock in winter. Roman farmers terraced their fields to reduce soil erosion, limed and fertilized their soils with livestock manure and monitored soil quality based on color, smell, taste, stickiness and compactness.

When Europeans arrived on the American continent, it was fully settled by Native Americans using foraging as well as all three systems of early agriculture.[7] In the eastern forests, some Native Americans depended mostly on foraging and others on horticulture, but many used both to their advantage. The Midwest and Great Plains were home to sedentary agriculturalists who managed crops in floodplains and hunted buffalo moving through their lands, as well as pastoralists who followed the buffalo throughout the year. Foragers, pastoralists, horticulturalists and sedentary agriculturalists all called the Far West home. Throughout the continent, these three early agricultural strategies, supplemented with foraging and trade, were used to create local, place-based ecological solutions to the problem of food acquisition.

1.1. Native American Farming Cultures

Prior to European contact, Native Americans farmed using sophisticated systems of crop production involving common agricultural practices such as irrigation, terracing and crop rotation, fertilization, and breeding. The Mohawk, Cherokee, Mandan and Hohokam are representative of farming cultures in the North American Northeast, Southeast, Northern Great Plains and Southwest.

The Mohawk inhabited much of what is now New York State. These shifting horticulturalists lived in villages, each featuring a large communal-style longhouse made from wood harvested while land was cleared for crop production. Mohawk women farmed forest gardens planted with the staple crops of corn, squash and beans—known as the "three sisters"—as well as pumpkins and tobacco. They processed crops and stored them for use over winter, gathered wild fruits and other plants, and tapped sugar maples for syrup. Mohawk men hunted deer and other wild game and fished to supplement crop production. When soil nutrient depletion caused crop production to decline in the forest garden, the village would move to a new location and establish a new garden and longhouse.[8]

The Cherokee originally inhabited a large region of the southern Appalachians including the Carolinas, northern Georgia and Alabama, southwest Virginia and the Cumberland Basin of Tennessee and Kentucky. They settled along major rivers in permanent villages and towns of homes and public buildings constructed with wood and stone and protected by wooden palisades. Sedentary agriculturalists, the Cherokee tended large communal fields of corn and beans on the outskirts of each village. These fields were managed

FIGURE 1.2. A Mandan Farmer Cultivating Maize and Squash with a Bone Hoe.
Credit: Clark Wissler

in a three-year rotation that included one year of corn and beans followed by two years of fallow. Each household also grew sunflowers, pumpkins and squash in smaller gardens located in fertile floodplains. The harvests from communal fields were shared with every family in the village. To supplement crop production, Cherokee women foraged wild fruits and other plants and men hunted deer and small game in the forested woodlands.[9]

The Mandan lived in permanent settlements along the banks of the Missouri River and two of its tributaries in present-day North and South Dakota. Their villages featured large, round, earthen lodges some forty feet in diameter surrounding a central plaza. The Mandan farmed, gathered, hunted and actively traded goods with other Great Plains tribes. Mandan women farmed small fields of corn, beans, sunflowers, tobacco, pumpkins and squash in fertile floodplain soils. Harvests were processed and stored for winter use and for trade. Mandan men hunted bison, deer and small mammals to provide meat and other material goods.[10]

The Hohokam lived along the Gila and Salt Rivers systems in what is now Arizona. One of several Southern farming peoples, they are distinguished by their construction of a major system of irrigation canals that spread over several hundred miles. The Hohokam used the canals to irrigate fields of corn, squash, beans and cotton. This irrigated agriculture was highly productive, allowing the harvest of two or sometimes three crops on the same land each year. Hunting wild game and gathering wild fruits and other plants provided important additions to their diet. Crops were processed and stored for use during the long dry periods each year when they could not be cultivated. The Hohokam and other Southwest farming cultures developed other innovations to increase agricultural yields. They developed improved corn and bean varieties that were well adapted to the extreme growing conditions typical of their region and used water conservation practices like terraces and check dams to capture and store water during heavy rains or flooding, and to trap silt to create deep fertile pockets of soil in the landscape.[11]

European colonists settling in North America faced the same ecological conditions as Native Americans, but brought with them the knowledge, experience and technology of intensive sedentary agriculture as practiced in Northern Europe. They faced three key challenges when they arrived on the eastern coast of America: climate, forests and soils. The climate was mostly warmer and wetter that their homelands, the land was predominately forested, and the soils were infertile. A blending of New World and Old World agricultural knowledge offered the best solution to the food acquisition problem.[12]

Europeans settling in the eastern forests practiced *horticulturalism* to create a distinctly new form of American agriculture. They usually cleared about 25 percent of their land and left the rest in forest. A typical rotation of the time was corn, wheat and pasture, with the length of each grain crop determined by corn yield. When corn yields fell, typically after two or three years, wheat was grown until yields fell, and then cultivated fields were planted in pasture or allowed to revert back to forest. Corn came first in the rotation because high grain yields could be produced in the rough ground of newly cleared fields by planting corn in hills, along with beans and squash. Cattle and hogs free-ranged in the forests and were rotated through pastured areas. Milk cows were typically kept close to the house and taken out to graze every day in improved pastures held in common. Cattle and hogs were fed some grain when there was excess, particularly right before slaughter.

Using this blend of Old and New World agriculture, eastern farms produced enough food for the family, plus some excess to barter or sell. Wheat, cattle and hogs were the principal commercial products, along with small quantities of butter and cheese made from excess milk produced by the family cow. Colonists bartered for or purchased specialty foods like sugar and salt, as well as manufactured goods they could not produce on the homestead. This system was productive for a time, but soils were soon degraded by the continuous production of grains, the low quantities of manure applied to cultivated fields and the short period of forest fallow.

As New Englanders pushed west over the Appalachian Mountains looking for new land, they took this new American agriculture with them and settled into the more fertile soils of the Midwest. Corn remained the grain of choice for human and livestock consumption. Because of limited transport options and prevailing market demands, wheat, beef and pork remained the most common commercial products. As cheaper Midwestern grain and livestock began to flow to markets to the east, farmers in the East switched to more perishable products like fruits, vegetables and dairy to remain profitable and to supply expanding urban markets nearby.

While small family farms were typical of agriculture in the Northeast and Midwest, an early form of industrial agriculture, powered by human labor, shaped the culture of the South. Most southern farms were family farms of less than 100 acres producing subsistence crops and livestock plus some tobacco and cotton for trade; however, large plantations producing commodity crops for export on 1,500 to 2,000 acres dominated commercial agriculture in the South. During the early colonial period, plantations in Maryland and points south along the Atlantic coast produced tobacco, indigo and rice for European markets, using first indentured servants and then African slaves. By the late 1700s, the Industrial

Revolution in England created a new market for American cotton. Southern planters, short on land and labor, pushed over the southern Appalachians to settle on new lands in the lower Mississippi region. By the early 1800s, more than one million African slaves worked the cotton, rice, tobacco and cornfields of the South in a system of industrial agriculture that dominated the South until the Civil War.

The western expansion of America through the mid-1800s offered both challenge and opportunity to agriculture. Climatic conditions in the Great Plains and arid West forced settlers to replace corn with more drought-resistant grains like winter wheat and sorghum, and swine with sheep and cattle. Even with these adjustments, conditions remained challenging, and sedentary agriculture in the region was very difficult. Many settlers found that pastoralism was better adapted to the climate of the American West and turned to raising cattle on the western prairies. Settlers who moved on to the Pacific Northwest and California found the moderate climate and fertile soils of the region excellent for the production of wheat and had developed a thriving industry exporting the grain to national and international markets by the 1850s.

The Rise of Industrial Agriculture

Industrial agriculture began to take its modern form in the mid-nineteenth-century wheat fields of California. The moderate Mediterranean climate, large flat valleys, plentiful water supplies, isolation from large population centers and access to major seaports set the stage for the continued development of mechanized commodity agriculture in America.

California agricultural investors understood that to be competitive in distant markets, they had to keep production costs low and provide consistent, high-quality products. The topography and climate of California's Central Valley facilitated the development of large-scale mechanization. Continuous waves of immigration supplied the low-cost hired labor needed on these large commercial farms, and San Francisco offered low-cost wind-powered transport to destinations worldwide. Wheat producers invested in affiliated industries—agricultural inputs, processing and marketing—and a vertically integrated, export-oriented commodity wheat industry emerged in the the Central Valley of California by the mid-1860s. This form of industrial production was fundamentally different from the small-scale, labor-intensive, owner-operated farms producing wheat in other parts of the country.

1.2. California's Central Valley

A unique mix of landscape, natural resources and climate support the incredible productive capacity of California's Central Valley, one of the most productive agricultural regions on Earth. Stretching over 450 miles from Redding in the north to Bakersfield in the south and covering about twenty thousand square miles, the valley produces a wide range of crops—more than two hundred and fifty at last count—and is home to nine of the top ten most productive agricultural counties in the nation.[13] The region's Mediterranean climate (hot, dry summers and cool, wet winters), a three-hundred-day growing season, rich floodplain soils, a level landscape and abundant water, plus easy access to national and global markets, allowed for the development of the most intensive agricultural production in the United States. Using less that 1 percent of the nation's farmland, Central Valley farmers produce 25 percent of the nation's food supply and 40 percent of its fruits, nuts and other table foods.[14]

The Central Valley is made up of two different valley systems that meet just east of San Francisco: the Sacramento Valley to the north and the San Joaquin Valley to the south. The Sacramento Valley is cooler and wetter with less productive soils but more abundant groundwater and more surface water resources, which it shares with the much drier and warmer San Joaquin Valley through a system of federal and state water projects.

Agriculture is threatened throughout the Central Valley by development pressures, more frequent droughts, reduced winter snowpack and increased competition for water. However, the more productive San Joaquin faces the most critical challenges. Groundwater levels in much of the region have dropped 120 feet or more in the last 150 years,[15] increasing the costs of pumping groundwater, causing massive land subsidence in some areas during the twentieth century (Figure 2)[16] and possibly increasing the risk of earthquakes along the San Andreas Fault.[17] More than half a million acres of cropland have been degraded by the buildup of salts from irrigation waters,[18] another two hundred thousand acres are challenged by drainage issues, and farms are competing intensely for imported surface water supplies with urban, industrial and environmental management uses. Some in the region are calling for the retirement of land in agricultural production as a sustainable solution.[19]

In 1870, California was a leading US producer of wheat and barley and was exporting these grains to Australia, China and Great Britain. By 1890, two decades of continuous grain production with little use of crop rotation, fertilizers or fallow had degraded soils and promoted high weed populations. Grain yield and quality had also begun to decline, and by the end of the century, California had abandoned wheat as a commercial crop. But the industrial structure created to produce the grain was easily adapted to other high-value crops. By 1910, California industrial agriculture had shifted focus to fruit and vegetable production, invested in the necessary irrigation infrastructure and developed the associated industries to pack, process and transport fresh and processed fruits and vegetables to national and international markets.

The Pacific Northwest climate and landscape was also well-suited to grain, fruit and vegetable production, but lacked the transport and irrigation infrastructure to develop a productive commercial industry. Railroads reached the region in the 1880s, and irrigation shortly thereafter, and a commercial agriculture sector swiftly developed in the region.

By 1920, California and Washington led the nation in fruit production. Washington was the number one apple producer, a position it holds to this day. California contributed significant quantities of apples, pears, cherries, peaches to the domestic supply and produced virtually all of the almonds, apricots, walnuts, olives and lemons grown in the United States, as well as substantial proportions of grapes and figs (80 percent), prunes and plums (70 percent) and oranges (57 percent).

Significant agricultural developments in the rest of the United States during the latter part of the 1800s encouraged the development of the fed cattle industry—a third major industrialization of American agriculture. The settling of the Great Plains and the mechanization of grain production increased wheat and corn supplies, while the invention of refrigerated railcars encouraged cattle feeding facilities and allied industries—slaughter, packing and distribution—to locate near major railheads in Chicago and Kansas City. The first grain distribution companies formed at this time. Initially these companies simply managed grain supplies, but they soon began to market US grain internationally and eventually developed into an industrial food-processing sector. By the end of the 1800s, the Midwest had emerged as the nation's leading corn and swine producer, while agriculture in the East adjusted to new competition from California fruits by shifting to fresh vegetables and dairy for regional consumption.

All of the best agricultural land in the United States was settled by 1920, closing the option of moving west as a way to increase production. Growing international demand drove US agricultural development through the remainder of the twentieth century. The solution to the need for increased production was to intensify agriculture—to produce more, on less land, with less labor. Over the next fifty years or so, with the help of a national system of publicly-funded land grant colleges, agricultural research stations, the Cooperative Extension Service and American agribusiness, American agriculture fully industrialized.

This process involved a massive transformation—from human and animal power to fossil fuel power, from a focus on subsistence to a focus on commercial production, from pasture-based livestock production to confined livestock production and from a regional market orientation to an international market orientation. During this period—roughly the middle of the twentieth century—the responsibility for the American food supply was transferred from a multitude of small-scale, diversified family farms and ranches to a small number of large-scale, highly specialized industrial farms and feedlots. In the eastern states, farming declined because of development pressure and because the region's small farms could not compete with foods produced in the West, where access to cheap energy, irrigation, new technologies and a national transport system made possible through public subsidy reduced the costs of production and increased yields on millions of acres.

Industrial agriculture solved the problem of managing land, labor, water, crops and livestock to produce food in a completely revolutionary way—by replacing labor and land with fossil fuels and abandoning subsistence as a fundamental purpose of agriculture.[20] Over the last century, the industrialization of the US food system drove a consolidation and regional specialization of the agriculture that continues to this day.

The Geography of the US Food Supply

America's plate is filled to overflowing with industrial food produced at home and sourced around the world.[21] California leads the nation in food production and processing, but specific areas within the Pacific Northwest, the Midwest, the Great Plains and the Southeast contribute significant volumes of just one or two kinds of products to the US food supply. Imports, primarily from Mexico, Canada, Chile, Central America and Asia, fill seasonal gaps in domestic production or provide a type or quality of product not produced domestically such as fresh vegetables in winter or tropical fruits year-round. Trade in food and other agricultural products has been important to America since colonial times, but the global movement of food, particularly perishable produce, has increased dramatically since the early 1980s, when the necessary technology and transportation systems were developed (see Figure 1.3).[22]

Fruits and Vegetables

California has long been the nation's leading producer of fruits, nuts and vegetables. With ample supplies of water for irrigation, California's ideal climate has supported an incredibly diverse and productive agricultural industry that has come to dominate US fruit, nut and vegetable production. Over the last decade, California has produced about 90 percent of all nuts, 70 percent of all processed vegetables and 50 percent of all fresh vegetables grown in the US each year. Other leading fresh vegetable producing states, Florida, Arizona, Georgia and Washington, together with California accounted for about 75 percent of US domestic fresh

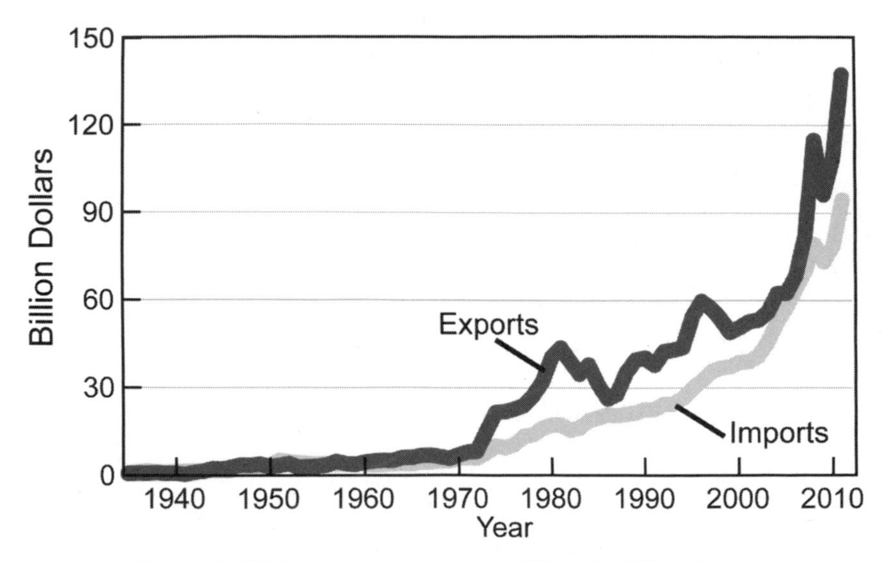

FIGURE 1.3. Trends in US Agricultural Imports and Exports. US trade began to grow after 1980 as changes in technology and trade policies opened markets and made it possible to transport perishable goods long distances without a significant loss of quality. Climate is an important factor in the global production and movement of food and agricultural materials. Through seasonal weather effects and other associated impacts, climate change will increase uncertainty in the global food system. The graph shows US imports and exports during the period 1935–2011 in adjusted dollar values. Credit: USGCRP

vegetable production in 2012. About 25 percent of the vegetables consumed in the US are imported, mostly from Mexico and Canada.

Together, three Pacific states—California, Washington and Oregon—produce much of the US-grown fruit supply: virtually all of the apricots, avocados, nectarines, pears, boysenberries and raspberries; more than 95 percent of the nation's strawberries and sweet cherries; more than 75 percent of the apples, prunes and plums; 70 percent of the peaches and nearly half of all US-grown blueberries. Imports contribute almost 50 percent of the US consumption of fresh fruits, with Mexico, Chile and Costa Rica the major suppliers.[23]

Grains and Beans

Wheat is the principal food grain produced in the United States and is the third-largest grain crop by value, behind corn and soybeans.[24] Different types of wheat are grown in different regions of the United States. Bread wheat accounts for almost 70 percent of total US wheat production and is grown in the Great Plains. Kansas, Oklahoma and Texas together account for about 75 percent of total US production of bread wheat in 2012. Wheat used for cakes and cookies accounts for about 20 percent of total US wheat production and is produced east of the Mississippi River. The white wheat used to make pasta and cereal makes up the remainder of the US total and is grown in the Pacific Northwest, Montana and North Dakota. The United States is a major supplier of wheat and wheat products to global markets and has held the largest share in global trade for many decades. About half of the US wheat crop is exported each year.

The production of corn and soybeans, the nation's top two grain crops by value, is centered in the Midwest, with Iowa leading the nation in production of both crops in 2012. Although only 2 percent of the corn produced in the US is directly eaten by people, it is an important livestock feed and is also processed into a multitude of food and industrial products including starch, sweeteners, corn oil, beverage and industrial alcohol and fuel ethanol. In 2012, about 8 percent of US production was exported, mostly for use as livestock feed. Soybeans are crushed to produce soybean meal and soybean oil. The meal is used as livestock feed while soybean oil is used for human consumption, accounting for more than half of all the vegetable oil consumed in the United States.[25] The United States is the world's largest exporter of soybeans, exporting about half of the US grown supply each year.

US rice production accounts for less than 2 percent of the world total, but it is important in global trade because most rice is consumed where it is produced. In recent years, about half of US production has been exported. Arkansas is the nation's leading producer of rice, accounting for about 50 percent of US production and together with Louisiana, Mississippi, Missouri and Texas produced almost 80 percent of US rice production in 2012.

Dry edible beans are produced in the Pacific Northwest and the Northern Great Plains. Pinto, navy, black, garbanzo and red kidney are the principal dry beans produced in the United States, with North Dakota leading the nation in production for more than 20 years. In 2012,

North Dakota, Idaho and Washington produced more than 50 percent of the total US production of dry edible beans.

Meat, Seafood and Dairy

America's beef production is centered in the south-central and western Great Plains. This region is home to three of the top five calf-producing states and processes about 70 percent of all US-grown beef. Although calves are produced on farms and ranches throughout the United States, most of them spend the last months of their lives in feedlots in Texas, Nebraska and Kansas, which together produce nearly 60 percent of the nation's fed cattle. About 12 percent of the US beef supply is imported as meat and livestock to assure a continuous supply of fresh beef throughout the year. Canadian imports contribute to the total US supply of grain-fed cattle, while Mexico supplies calves for US feeding operations during seasonal lows in domestic calf production.

US pork production is centered in the Midwest, home to nine of the top ten hog-producing states that together produced about 70 percent of total US pork production over the last decade. Iowa is the hog-producing capital of the nation, producing about a quarter of the country's pork each year. Although pork imports amount to just 4 percent of domestic consumption, Canada is a large supplier of feeder pigs to the United States, sending more than 5.6 million our way in 2012. The United States is the largest exporter of pork and pork products globally, with exports averaging 20 percent of commercial pork production over the last decade.[26]

American poultry production is centered in the Southeast. Georgia is the nation's leading producer of chickens and along with Arkansas, Alabama, North Carolina, Mississippi and Texas produces about two-thirds of the nation's supply. The United States is the world's largest turkey producer and largest exporter of turkey products. Turkey production is centered in the Midwest and the Southeast, with each region accounting for about 40 percent of national production in 2012. The United States is the world's largest producer and second-largest exporter of poultry meat, exporting about 20 percent of US poultry meat and 14 percent of turkey meat in 2012.

While the United States' seafood consumption is far behind that of poultry, pork and beef, it adds up to nearly five billion pounds each year, second only to China.[27] In 2011, about half the seafood consumed in the US was wild-caught, and the other half farmed. About 9 percent was produced domestically and the rest imported, mostly from Asia but also from Chile, the European Union and Canada. Asia accounted for 89 percent of world aquaculture production by volume in 2010, with China by far the largest producer.

The United States is the single largest producer of cow's milk in the world. About half of the milk supply is processed into cheese, another third into fluid milk and cream products and the remainder into dairy products such as butter, ice cream and milk powders. California and Wisconsin have led the nation in dairy production since the 1980s. For

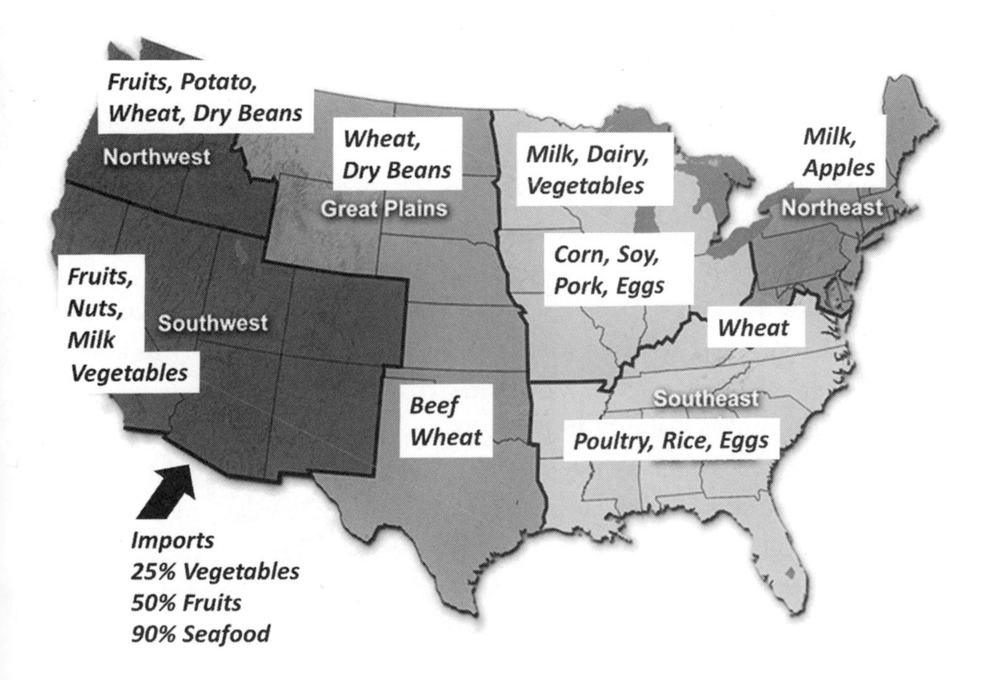

FIGURE 1.4. The Geography of Food Production in the United States. The regional specialization of food production emerged as a result of natural resource capacity and public investment in built resources such as irrigation and transportation networks. Regional production characteristics based on the 2012 Agricultural Atlas and USDA-ERS 2012 production data. Map Base Source: USGCRP

the last decade, they have together produced about one-third and one-half of the national supply of milk and cheese, respectively. Milk and dairy product imports to the United States are dominated by specialty cheeses from Europe and milk powders from New Zealand. About 13 percent of the US milk supply was exported in 2012.

Sustainability Challenges to the US Food System

Over the last century, the US food system has evolved to rely on large-scale, vertically integrated monocultures to produce, process and deliver food and other agricultural products to consumers in the United States and beyond. The success of the US industrial food system has been achieved through a focus on land- and labor-efficient production of commodities in a national system characterized by monocultures, geographic specialization and increasing concentration in all phases of the food system. For example, since 2000, oligopolies (4-firm control of at least 50 percent of market) have emerged in the US food system in agricultural inputs,[28] grain distribution,[29] food processing[30] and grocery retailing.[31] In 2007, just 2 percent of US farms and ranches accounted for more than 50 percent of all US agricultural sales and from 60 to 70 percent of the sales of high-value crops (vegetables, fruits, nuts, nursery and greenhouse products), hogs, dairy, poultry and beef.[32]

This industrial system of food production solved the age-old resource allocation problem for American consumers by offering those with the financial means a diverse selection of low-cost foods of consistent quality and year-round availability. Ironically, we seem to have returned to our ancestral roots—while we used to forage in the forests and grasslands of the world, now we forage in the supermarket. In both cases, a vast biological or bio-industrial machinery is at work behind the scenes. While the industrial food system operates out of sight and out of mind for most Americans, the simplicity of foraging at the local full-service grocery masks a variety of environmental and social harms that are increasingly undermining the sustainability of the US food system.

The environmental, social and economic harms produced by the US industrial food system have raised concerns since the early days of mechanization in the mid-1800s and are widely recognized today.[33] The linked 21st-century challenges of climate change and resource scarcity bring a new urgency to concerns about the sustainability of the US industrial food system.[34] The resource degradation associated with industrial production practices coupled with the geographic specialization and concentration of American agriculture degrade the nation's adaptive capacity, while the interactions between energy, water and land use in the US will likely amplify climate change effects[35] and significantly increase the vulnerability of the US industrial food system to climate change. Although the application of vulnerability assessment and adaptation planning to US agriculture and food systems is only just now getting underway, sustainable agriculture has been widely recognized as a promising strategy for the development of integrated solutions to the climate change challenges ahead.[36]

1.3. The Interaction of Water, Energy and Agriculture Amplifies Climate Change Effects[37]

The extraordinary drought and heat waves that spread across the United States in 2011 and 2012 hit particularly hard in Texas. The summer of 2011 was the warmest and driest summer on record (Figure 1.5). The associated heat wave kept temperatures above 100 degrees Fahrenheit for more than a month. These extreme climatic events set off a cascade of interacting effects among the region's energy, water and land resources.

High temperatures increased demand for cooling, which increased water withdrawals for electricity generation. Heat, increased evaporation, drier soils and drought also led to higher water demands, this time for irrigation, placing further stress on dwindling supplies. At the same time, low-flowing and warmer rivers could not provide enough cooling water for the power companies to supply the higher demands for electricity from consumers. The impacts on land resources and land use were dramatic. The drought cost Texas farmers and ranchers about a quarter of their normal income. Increased tree mortality in the region provided fuel for wildfires that burned

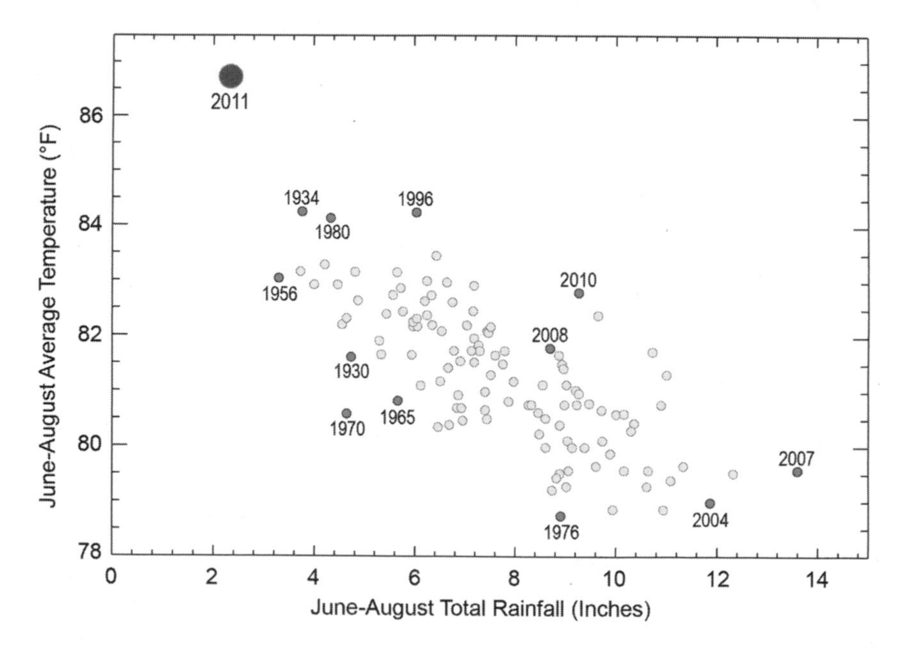

FIGURE 1.5. Texas Summer 2011: Record Heat and Drought. The dots on this graph show average summer temperature and total summer rainfall in Texas from 1919 through 2012. The dark dots illustrate the extremes in temperatures and rainfall observed over this time period. The record temperatures and drought during the summer of 2011 (large dark dot at upper left) represent conditions far outside those that have occurred since the instrumental weather records began. An analysis has shown that the probability of such an event has more than doubled as a result of human-induced climate change. Credit: USGCRP.

3.8 million acres and destroyed nearly three thousand homes. Because water shortages threatened more than 3,000 megawatts of generating capacity—enough to power one million homes—water was rationed to farms and urban areas to preserve water for energy production.

Sustainable Agriculture: A Resilient Agriculture?

New demands from a dynamic global economy, continued decline in the quality and availability of natural resources, and the unprecedented challenges of climate change are just beginning to take their toll on the US food system. While the challenge is daunting, we have a wealth of knowledge and experience to begin working on a new, more resilient solution to the age-old resource allocation problem. Much of this knowledge can be found in the principles and practices of sustainable agriculture.[38]

Compared to industrial production systems, sustainable production systems tend to enhance the resilience of the food systems within which they reside. Sustainable farmers and ranchers typically manage much greater biodiversity, employ more people in better jobs and circulate more dollars in the regional economy.[39] The diversified production systems typical of sustainable farms and ranches reduce the need for fossil fuel energy, water, pesticides and synthetic fertilizers and typically produce less waste.[40] Sustainable farms and ranches have high quality soils and tend to enhance environmental well-being in the communities where they are located.[41] Sustainable agriculture also offers an unprecedented opportunity to mitigate climate change while increasing agricultural productivity worldwide through the use of regenerative agricultural practices.[42]

There are successful sustainable producers engaged in food systems all across the US. These farmers and ranchers offer models of locally adapted food production, processing and distribution that cultivate ecological, social and economic well-being on their farms and ranches and in their communities. They have been successful largely without the scientific, technical and financial support provided to producers using industrial methods. Throughout the ecological and social diversity of the American landscape, on farms large, medium, and small, sustainable producers and their supporters have been busy for decades laying the foundation for a more sustainable and resilient American food system.

How are sustainable producers managing the challenges of farming in a changing climate? What kinds of changes have they perceived and how are they adapting to those changes? What practices do they use to cultivate resilience to climate change on their farms and ranches and in their communities?

1.4. Defining Sustainable Agriculture

Congress defined sustainable agriculture more than twenty years ago in the Food, Agriculture, Conservation, and Trade Act of 1990:

- the term sustainable agriculture means an integrated system of plant and animal production practices having a site-specific application that will, over the long term:
- satisfy human food and fiber needs
- enhance the environmental quality and the natural resource base upon which the agricultural economy depends
- make the most efficient use of nonrenewable resources and on-farm resources and integrate, where appropriate, natural biological cycles and controls
- sustain the economic viability of farm operations
- enhance the quality of life for farmers and society as a whole

This definition explicitly acknowledges the multiple dimensions of sustainability—ecological, social and economic—and provides specific design criteria for sustainable agriculture systems.

The focus on ecological health and community well-being as the basis of agricultural productivity clarifies a fundamental difference of philosophy between sustainable and industrial agriculturalists.[43] The industrial philosophy of agriculture views the production of agricultural products as no different from the industrial production of other goods. The farm is a factory, the land an assembly line. The farmer is a factory worker, managing purchased energy and materials to produce high volumes of low-cost commodities. A narrow focus on economic efficiency drives increasing geographic specialization, concentration and vertical integration in the US food system.

The agrarian philosophy of agriculture takes a broader view of the potential contributions of agriculture to society. This view recognizes that agriculture has the capacity to produce many ecological and social goods in addition to supplying sufficient food and fiber. Agriculture has the capacity to restore and conserve natural resources like soil, water and air. Healthy soils can help to replenish groundwater supplies and clean up surface waters, and crop diversity can reduce the need for fertilizers and pesticides, which are the largest source of water pollution in the United States. Rural communities benefit from the social and financial capital created by a focus on producing high-quality goods for regional markets. Urban communities benefit from the social and financial capital produced by regionally owned production, processing, distribution and marketing, as well as better access to nutrient-dense foods and improved environmental quality.

Voices from the Field

From the summer of 2013 through late winter 2014, I spoke about these issues with twenty-five award-winning sustainable producers from across the United States. We talked about some of the most challenging aspects of sustainable management on their farm or ranch, and about changes in the weather. The producers shared with me how they think about managing for resilience to climate effects and how confident they are that they can manage those effects in the coming years. Many also shared their concerns about the path the US food system has taken over the last fifty years and their frustrations with the scientific, economic, regulatory and policy barriers to sustainable food in this country.

All of the producers I interviewed have been farming in the same location for at least twenty years, many for thirty and some for forty years or more. Many are the third or fourth generation on the farm or ranch they now own and operate, so they have benefit of family memories and stories of weather long ago to inform their perceptions of the weather they have experienced through the years.

These farmers, growers and ranchers own successful agricultural businesses that operate, for the most part, outside of the industrial food system. Most have never participated in federal agricultural subsidy, production insurance, disaster or conservation programs. Most have been nationally recognized for their excellence in sustainable farm and ranch management. All are innovative leaders and compelling advocates for sustainable agriculture and food systems.

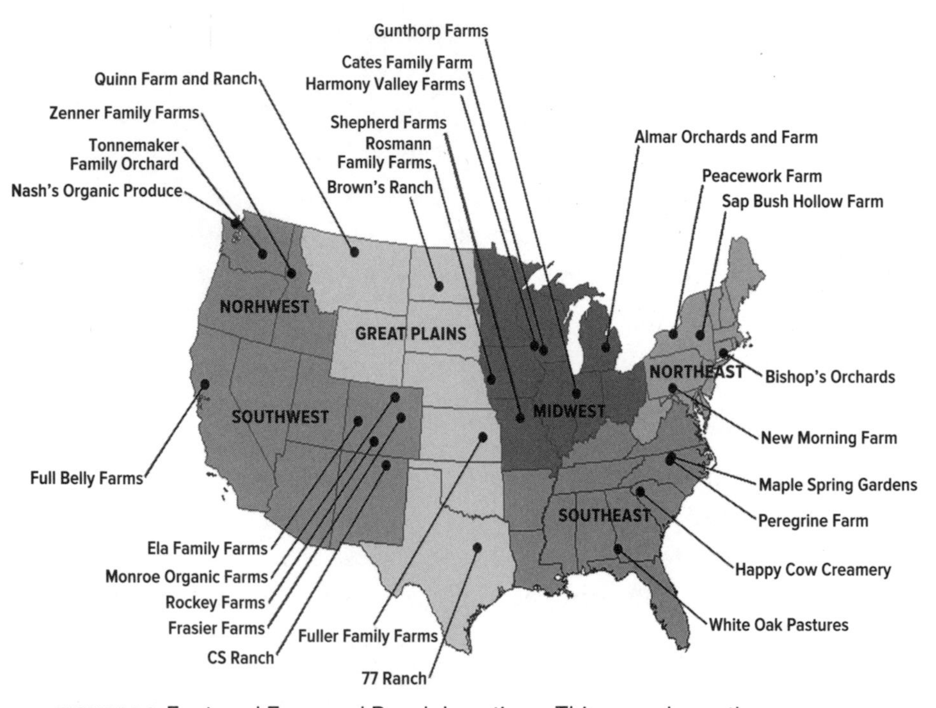

FIGURE 1.6. Featured Farm and Ranch Locations. This map shows the name and location of the farms and ranches owned by the 25 sustainable producers who contributed to this book through their participation in an hour-long semi-structured telephone interview conducted July 2013 to March 2014. Credit: Cherie Southwick and David Abernathy, GIS Crew, Warren Wilson College, Asheville, NC.

Vegetables, Fruit, Pork, Poultry and Grain in the Northwest

Farmers in the Northwest have enjoyed relatively moderate climate change effects over the last two decades, although warmer winters, declining snow pack and wetter springs and falls have complicated vegetable and grain production.

Nash Huber produces organic vegetables and fruits, food and feed grains, pork and poultry, and cover crop seed on 450 acres of irrigated prime farmland in Washington's North Olympic Peninsula. Warmer winters, wetter falls, more unpredictable weather in summer and more variable extremes have challenged farm management since about 1995. In response, Nash has purchased more tractors, tools, processing equipment and combines to take advantage of increasingly narrow windows of time when conditions are right for fieldwork.

Kole Tonnemaker and his brother Kurt are the third generation to own and operate the 126-acre Tonnemaker Hill Farm, home to 60 acres of orchards, 20 acres of vegetables and 40 acres of hay, all produced under irrigation and all certified organic. Although Kole can't say that he has seen a big change in seasonal weather patterns, in the early 1990s, he decided to add vegetables to the farm's crop mix to buffer farm income from the risk of fruit crop failure in the region's highly variable spring weather.

Russ Zenner has been farming in the Palouse Region of Idaho near the Washington–Idaho border in Genesee for more than 40 years. Russ manages 2,800 acres of dry-land direct-seeded crops in a three-year rotation of winter wheat, spring grains and spring broadleaf crops. Recent changes in seasonal rainfall patterns have got Russ thinking about redesigning the crop rotation to increase crop diversity and the proportion of fall-seeded broadleaf crops.

Vegetables, Fruit, Beef, Lamb and Grain in the Southwest

Many farmers and ranchers in the Southwest region have experienced changes in weather that have increased the challenges of maintaining the productivity of their farms and ranches.

Paul Muller co-owns and operates Full Belly Farm, a 400-acre diversified organic farm located in the Capay Valley of Northern California where more than 80 different crops including vegetables, herbs, nuts, flowers, fruits, grains and livestock are raised on a diverse landscape mosaic. Although water conservation has always been a management focus, heavier rainfall, longer dry periods and continuing drought have encouraged more thinking about sustainable water management. Changes underway include a switch to more drought-tolerant cover crops, the addition of cover-crop mulches to conserve soil moisture and upgrading to more water-efficient drip and microsprinkler irrigation systems.

Jacquie Monroe and her husband Jerry are the third generation to own and manage Monroe Family Farms, located east of Denver. The 200-acre farm produces 100 different vegetables and all the pasture, hay and feed grains needed to produce pasture-based meats (beef, pork and lamb) and eggs on the farm. Changing weather patterns have reduced water supplies, requiring investment in more efficient irrigation systems and a plan for adjusting crop production area to fit projected seasonal water availability.

Steve Ela is a fourth-generation fruit grower at Ela Family Farms located near Hotchkiss in the "frost-free" fruit-growing region of Colorado's Western Slope. Steve manages 100 certified organic acres planted with 23 varieties of apples and 29 varieties of pears, peaches, cherries, plums and tomatoes. More variable weather, reduced snowpack, more extreme weather and a lengthening growing season have required some significant changes in production practices to maintain farm productivity and profitability. Steve has installed more efficient drip irrigation, added new frost protection and has shifted to longer season apple varieties and more frost resistant peaches.

Brendon Rockey and his brother Sheldon are the third generation to grow potatoes on the 500-acre Rockey Farms, located in the San Luis Valley of Colorado. Dwindling water supplies and failing groundwater levels motivated Brendon to experiment with cover crops and companion planting to try and reduce water use. Brendon improved farm profitability with the new cropping practices through a 50 percent reduction in water use and decreasing or eliminating fertilizer and pesticide applications while maintaining yields and improving crop quality. Brendon can't say that the current drought is a sign of climate change; he is just working to make his farm as resilient as possible to whatever the future may bring.

Mark Frasier has managed the 29,000 acre Woodrow division of Frasier Farms for more than 30 years. The third generation on the ranch, Mark buys stocker cattle and raises calves produced on the ranch to run nearly 5,000 head when fully stocked. Mark has not noticed a change in weather patterns during his lifetime, but credits planned grazing with the resilience of his operation to the weather extremes typical in the short grass prairie region of eastern Colorado.

Julia Davis Stafford is a fourth-generation co-owner and operator of the CS Ranch, located in northeastern New Mexico on 130,000 acres. Julia has worked together with her family to manage the ranch's cow-calf and stocker enterprises for more than 30 years. For much of the ranch's history, the cowherd has numbered between 2,500 and 3,000 head, but continuing drought over the last 15 years has forced the family to destock the ranch. Julia is concerned that eventually the ranch will be unable to maintain profitability if cattle numbers continue to fall, creating widespread impacts to the livelihoods of ranch owners, employees and local businesses.

Vegetable, Beef, Lamb, Poultry, Grain and Beans in the Great Plains

Farmers and ranchers in the Great Plains, a region of weather extremes, have been challenged by warmer temperatures, more extreme rainfall in the north and more extreme drought throughout the region.

Gabe Brown has been producing cattle, feed and food grains on 5,000 acres of native rangeland, perennial forages and no-till cropland near Bismarck, North Dakota, for more than 30 years. Gabe thinks the most effective climate risk management tool he has available is the capacity of the ranch's healthy soils to buffer the more variable rainfall and temperatures increasingly common today.

Bob Quinn owns and manages 4,000 acres of organic land producing certified organic food grains in a full tillage dryland production system near Big Sandy, Montana. More weather extremes and dry fall conditions have required some changes to his crop mix and fieldwork scheduling, but warmer winters have created new fruit growing opportunities.

Gail Fuller is the third generation to farm at Fuller Farms, a 2,000-acre family farm located in east-central Kansas near Emporia. Gail manages a large variety of cash crops, cattle, sheep and poultry in a highly diversified and integrated dryland production system. More variable weather encouraged Gail to increase the diversity of his grain rotation by adding cover crop cocktails so that his soils could hold more water between rains.

Gary Price uses planned grazing to manage a cow-calf and stocker operation on the restored native range that dominate the landscape on his 2,500-acre 77 Ranch, located near Blooming Grove, Texas. The ranch also includes 200 acres of cropland, 90 acres of improved pastures and

40 acres of small stock ponds and natural lakes. Gary has reduced herd size and replaced some grains with forage crops to provide supplemental feed for his cattle in response to extreme drought conditions.

Vegetables, Fruit, Nuts, Beef, Pork, Poultry and Grains in the Midwest

In a region well-known for extremes of weather, rapid temperature swings, more variable rainfall, more very heavy downpours and catastrophic flooding have challenged many Midwest producers.

Ron Rosmann produces certified organic feed and food grains, beef, pork and poultry on the 700-acre Rosmann Family Farms located in west-central Iowa. More frequent and intense weather extremes since about 2000 have required more careful livestock management during periods of extreme temperature swings, and there are new weed-management challenges in the grain fields that seem to be related to a lengthening growing season.

Dan Shepherd has grown pecans, buffalo, gamma grass seed and grains at Shepherd Farms near Clifton Hill in north-central Missouri, but now he manages about 300 acres of mature pecans on the 4,000-acre farm. Although Dan has not noticed any clear trends in changing weather patterns, the last decade has included several weather-related firsts in his 44 years on the farm: total crop loss from a spring freeze in 2007, levee breaches in 2008 and 2013, and a prolonged 3-year drought from 2011–2013.

Greg Gunthorp is a fourth-generation hog farmer who finishes cattle, swine and poultry on pasture at Gunthorp Farms near LaGrange, Indiana. He processes the beef, pork and poultry that he raises on-farm in a USDA-inspected facility that he built on the farm and sells it through an on-farm store and into local wholesale markets. Greg has not perceived any significant changes in weather during the twenty-five years he has managed the farm, other than perhaps a slight increase in extreme events.

Richard de Wilde owns Harmony Valley Farm, a 200-acre diversified vegetable farm located near Viroqua in southern Wisconsin that sells certified organic produce, berries and beef through direct and wholesale markets. He realized that there "is no normal anymore" after back-to-back, 2,000-year flood events ended the 2007 growing season two months early and destroyed early summer crops in 2008. He credits his local community and the farm's CSA members with providing crucial support during recovery from the catastrophic flooding.

Dick Cates and his wife, Kim, own Cates Family Farm, a 900-acre grass-fed beef farm near Spring Green in southern Wisconsin. In 1990 Dick transitioned the farm to rotational grazing practices to restore the farm's native oak savannah landscape and improve profitability. Over the last 10 to 15 years, Dick has reduced stocking rates, increased hay reserves and is thinking about adding an irrigation system to the farm in response to more variability and extremes in weather.

Jim Koan produces organic apples in a 150-acre orchard, as well as vegetables, grains, pasture-raised hogs and value-products on the 500-acre Almar Farm and Orchards in eastern Michigan. The fourth generation to own the farm, Jim has managed it for more than 40 years. Since about 2000, warmer winters and more variable springs have increased the risk of crop failure, and imported apples have taken a bite out of profits. Jim reduced production risk and improved profits by diversifying into value-added apple products, including apple-finished pork, apple cider vinegar and an award-winning hard cider.

Vegetables, Fruit, Beef, Pork, Lamb and Poultry in the Northeast

Some farmers in the Northeast have been challenged by more variable weather, heat waves, more heavy rainfall and extreme weather events and an increase in plant disease.

Jim Crawford has produced a diverse mix of organic vegetables and small fruits—about 50 different kinds—on 45 acres at New Morning Farm in south-central Pennsylvania for 44 years. About 15 years ago, more variable temperatures, more heavy rainfall, summer drought and a growing number of novel plant diseases began to complicate vegetable and fruit production on the farm. Jim has adapted by shifting soil preparation to the fall, protecting soils and crops from excessive rainfall with plastic mulch and hoophouses, and increasing use of OMRI-approved pesticides.

Elizabeth Henderson has grown organic vegetables at Peacework Organic CSA for more than 25 years near Newark, NY, on prime farmland under a conservation easement. Hotter summers, more heavy rainfall and drought, and novel diseases have required some adjustments to the farm's management practices, including the addition of irrigation, changes to fieldwork hours and adjustments to crop succession timing.

Jim Hayes has been raising livestock since 1979 at Sap Bush Hollow Farm west of Albany, New York. Jim produces beef, lamb, pork and poultry on 160 acres of pasture using an integrated rotational grazing system. More dry periods and drought, more frequent and stronger winds, and extreme weather have challenged farm operations and required new infrastructure such as drainage systems, a raised barn, reinforced pasture shelters and solar power.

Jonathan Bishop is a fourth-generation co-owner and field operations manager at Bishop's Orchards in Guilford, Connecticut, a 320-acre diversified fruit and vegetable farm supplying multiple direct markets, including a full-service grocery on the farm. Although Jonathan hasn't seen any long-term changes in weather patterns, a number of severe storms over the last few years have confirmed for Jonathan the resilience benefits of scale, experience and crop diversity.

Vegetables, Fruit, Milk, Beef, Pork, Lamb and Poultry in the Southeast

Farmers in the Southeast report increased challenges from more frequent summer droughts and heat waves and more frequent and intense weather extremes of all kinds.

Ken Dawson and his family have been producing vegetables, cut flowers, culinary and medicinal herb starts and small fruits since 1981 at Maple Spring Gardens, located northwest of Durham, NC. To maintain productivity in recent years, Ken has had to adapt his cropping practices to increasing extremes of weather and warmer summer temperatures. He has stopped growing some crops and has adjusted seasonal planting schedules to avoid new disease challenges and to achieve continuous vegetable harvests through the growing season.

Alex Hitt and his wife Betsy grow vegetables and flowers for wholesale and direct markets at Peregrine Farm located just west of Chapel Hill, North Carolina. More intense heat waves and drought in summer, combined with a reduced summer water supply, have got Alex thinking about putting more focus on vegetable production in the spring and fall and reducing crop production in mid-summer.

Tom Trantham owns and manages Happy Cow Creamery, a 90-cow grass-based dairy farm and creamery located south of Greenville, South Carolina. Dryer summers and more intense summer thunderstorms over the last decade have not presented any challenges on the farm, largely because Tom's use of annual forage crops in a diverse rotation makes it easy to adjust to changing weather conditions throughout the year.

Will Harris is the fourth generation to own and operate White Oak Pastures, a pasture-based livestock farm located about 90 miles from the Gulf of Mexico in southwest Georgia. Livestock produced on White Oak Pasture's 2,500 acres of non-irrigated pastures are processed on farm in state-of-the-art zero-waste USDA-inspected beef and poultry processing plants. Will thinks hotter summers, more extreme drought and competition for water have begun to challenge livestock management on his farm. If dry periods and droughts continue to intensify in coming years, Will believes he will have to find a way to irrigate his pastures in a region marked by increasing competition for limited groundwater resources.

The Climate Change Challenge

Agriculture has proven to be a robust food acquisition strategy with an incredible record of adaptation over more than five thousand years of human experience. During that long period of time, every time population pressures have intensified, we have found a way to increase food production in nearly every biome on the planet capable of supporting agriculture.

We have increased food production through two complementary strategies. We can grow more food by cultivating more acres, an *extensification* of agricultural production. The settlement of the Midwest and Great Plains was a major extensification of American agriculture. The production of grains and beef increased dramatically during this period even though the yields of crops and livestock per acre of land remained the same. Even today, American agriculture still makes use of an extensive land base to produce some products like beef calves, cattle and lamb.

The other strategy is to grow more food by increasing crop and livestock yields, an *intensification* of agricultural production. This approach uses land more intensively and requires major changes in agricultural management. These changes could involve the selection of high-yielding plants and animals, or working to prevent conditions that reduce yields like pests or disease. They may involve bringing additional resources, like water or nutrients or pesticides, to the land to address a condition that is limiting the growth of crops or livestock. Industrial agriculturalists have used this strategy to greatly increase production without adding to the agricultural land base through the development of tools and technologies that can produce a substantial increase in yield per acre, although often at a reduced energy profit: irrigation, improved crops and livestock, fertilizers, pesticides, antibiotics and confined animal production.

These two approaches worked well for us for a very long time. It is no wonder that many in the US industrial agricultural community—farmers, growers, ranchers, researchers, technical advisors, extension policymakers, agribusinesspeople—are upbeat about our capacity to adapt agriculture to climate change; however, their confidence may be misplaced.

Prime agricultural land is dwindling, and human population is growing, limiting our options for the further extensification of food production. Industrial agriculture degrades the ecological, economic and social foundations of human well-being and fuels global warming, limiting the intensification of production as a solution. And it is easy to forget that agriculture—all of agriculture, from its first beginnings in Mesopotamia ten thousand years ago to the satellite-controlled tractor fertilizing genetically engineered crops in Iowa last week—evolved during a period of unusual climatic stability and with the benefit of a wealth of human, natural and technological resources.

As we face the climate change challenge, we do not have the benefit of a stable climate. We do not have access to unlimited natural, financial, scientific or technological resources. The natural resource base upon which agriculture depends is oversubscribed, and much of it is highly degraded. And the very system that we depend on for food fuels the pace and intensity of climate change. As a species dependent on agriculture for our survival, we have entered uncharted territory.

Ten thousand years ago, in many parts of the world, our ancestors evolved from food foragers to shifting horticulturalists and pastoralists to sedentary farmers in response to a suite of challenges not unlike those we face today. Population pressures were building, natural resources were declining, and changing climatic conditions reduced the productivity of the food acquisition strategies our species had used for at least two hundred thousand years. Agriculture emerged as the most successful solution to the resource allocation problem, and agricultural peoples soon dominated the planet. We face an unprecedented challenge, but it is a challenge that our species has navigated at least once before. What kinds of knowledge, tools and experience can help guide us to a sustainable and resilient future?

Never before in our history as a species have we known so much about how our planet works or how much our own well-being depends on the well-being of healthy ecosystems. Never before have we had such an ability to envision a sustainable and resilient path into our changing world. How will we solve the problem of food acquisition in a world shaped by the unprecedented challenges of resource scarcity and climate change?

This is the climate change challenge.

1.5. Resource Scarcity Limits Industrial Solutions

Will Harris owns and operates White Oak Pastures, a pasture-based livestock farm in southwest Georgia near Bluffton. Will's father managed operations through the mid-1900s, and he changed with the times—shifting from pasture-finished to finishing cattle on purchased feeds—to keep White Oak Pastures up to date. Will continued those practices when he took over the farm in the 1970s, but became disenchanted with them and in 2000 began transitioning the farm to sustainable pasture-based livestock production.

Will shares these thoughts about the differences between industrial and sustainable agriculture:

"You know, most of the people that operate in this high animal welfare/high environmental sustainability farming model are younger people who came to it from somewhere else. Not many of them are like me, or only a small percentage—former monocultural, commoditized, centralized, industrialized farmers—but I'm one of them. And I'm like a reformed prostitute, you know, I've got the zeal of the convert.

"I love the debate that I get into occasionally, probably a little bit more than most, because I am one of the good old boys. They say, 'You know, the way you farm won't feed the world versus the way we farm. We're feeding the world,' and I love it when they say that, because they say, 'You just can't produce enough.'

"When they say that, I try not to smile, and I say, 'Okay, let's have that debate, but before we have that debate, I want us both to stipulate that neither farming system will feed an endlessly increasing population.' The Earth has got a carrying capacity, and once you get beyond that carrying capacity, neither one of them is going to feed the world.

"And most of them will stipulate that. They don't want to, but they eventually will. And I say, 'Okay, well, I'll go ahead then and capitulate right up front that if we're going to run out of acres first, you win. You can feed way more people than I can if acres are the only limiting factor. If we've got unlimited water, unlimited petrol fuel, unlimited antibiotics that don't create pathogen-resistance, unlimited fertilizer resources...you win.

"But now if the limiting factor becomes water, I'm probably going to win, because I don't use as much water as you do. If the limiting factor becomes petrol fuel, I win, because I don't use as much of it as you do. And if the limiting factors become phosphates and potash and these other depleting resources, I win, because I don't use as much as you. And antibiotics and pesticides, and so on...I win just about any way we do it other than acreage."

Endnotes

1. Part 1 of Ernest Schusky's 1989 book, *Culture and Agriculture: An Ecological Introduction to Traditional and Modern Farming Systems,* is a fascinating and thorough discussion of the evolution of agriculture in different parts of the world. Published by Greenwood Press, New York, NY.
2. Ibid.
3. Ibid.
4. J. Diamond. 1997. To Farm or Not to Farm, Ch. 6, in *Guns, Germs, and Steel: The Fates of Human Societies.* New York: Norton.
5. Ibid.
6. E. Schusky. 1989. *Culture and Agriculture: An Ecological Introduction to Traditional and Modern Farming Systems.* Greenwood Press: New York.
7. R. Hurt. 1994. *American Agriculture: A Brief History.* Iowa State University Press: Ames.
8. Ibid.
9. Early Cherokee Agriculture. ancientworlds.net/aw/Article/1202513
10. R. Hurt. 1994. *American Agriculture: A Brief History.* Iowa State University Press: Ames.
11. Ibid.
12. Unless otherwise noted, the section on US agricultural history is adapted from R. Hurt. 1994. *American Agriculture: A Brief History.* Iowa State University Press: Ames.
13. Tulare Agricultural Commission News Release. 2014. agcomm.co.tulare.ca.us/default/assets/File/2012CensusCA_1.pdf
14. California's Central Valley, USGS. 2014. ca.water.usgs.gov/projects/central-valley/about-central-valley.html
15. D. Galloway, D. Jones, and S. Ingebritsen. San Joaquin Valley, U.S. Geological Survey Circular 1182. pubs.usgs.gov/circ/circ1182/pdf/06SanJoaquinValley.pdf
16. Ibid.
17. C. Amos et al. 2014. Uplift and seismicity driven by groundwater depletion in central California. *Nature* 509, 483–486 (22 May 2014). nature.com/nature/journal/vaop/ncurrent/full/nature13275.html
18. J. Lund and T. Harter. 2013. California's Groundwater Problems and Prospects. California Water Blog. californiawaterblog.com/2013/01/30/californias-groundwater-problems-and-prospects/
19. For example, Why Retirement Makes Sense in the Westlands Water District westlandswater.org/long/200201/landretirebro.pdf
20. Thomas Lyson tells the story of the transformation of US agriculture from craft to commodity, including a discussion of the underlying economic

philosophy of industrialism, in his book *Civic Agriculture: Reconnecting Farm, Food and Community*. 2004. Tufts University Press: Lebanon, NH.

21. All of the information on specific crops and livestock production were obtained from the Agricultural Marketing Resource Center at agmrc.org unless otherwise noted. Production statistics included in this section were calculated using data from USDA 2012 Annual Production Summaries unless otherwise noted.
22. B. Halweil. 2002. Homegrown: A Case for Local Food. World Watch Paper 163.
23. S. Huang. 2013. Imports Contribute to Year-Round Fresh Fruit Availability. USDA-ERS FTS-356-01.
24. USDA-ERS Wheat Overview. 2013. ers.usda.gov/topics/crops/wheat.aspx#.U64NTLEuKQo
25. American Soybean Association. Domestic Utilization. Soystats 2012. soystats.com/archives/2012/non-frames.htm
26. USDA-FAS Livestock and Poultry World Markets and Trade. 2014. fas.usda.gov/data/livestock-and-poultry-world-markets-and-trade
27. Fish Watch. U.S. Seafood Facts. The Surprising Sources of Your Favorite Seafoods. 2013. National Oceanic and Atmospheric Administration. fishwatch.gov/features/top10seafoods_and_sources_10_10_12.html
28. K. Fuglie et al. 2012. Rising Concentration in Agricultural Input Industries Influences New Farm Technologies. Amber Waves, USDA-ERS.
29. Concentration in the Food Industry. 2014. CorpWatch. community.corpwatch.org/adm/pages/food_industry.php
30. Ibid.
31. Food and Water Watch. 2010. Consolidation and Buyer Power in the Grocery Industry Factsheet. documents.foodandwaterwatch.org/doc/RetailConcentration-web.pdf
32. R. Hoppe and D. Banker. 2010. The Structure and Finances of U.S. Farms. ERS Economic Information Bulletin # 66.
33. Some noteworthy examples include *Silent Spring* by Rachel Carson (1963), *Radical Agriculture*, edited by Richard Merrell (1976), *The Unsettling of America* by Wendell Berry (1977), *New Roots for Agriculture* by Wes Jackson (1980), *Alternative Agriculture* by the National Research Council (1989), *Sustainable Agriculture Systems*, edited by C. Edwards, R. Lal, P. Madden, R. Miller and G. House (1990), *From Land to Mouth: Understanding the Food System* by Brewster Kneen (1995), *Life Cycle-Based Sustainability Indicators for Assessment of the U.S. Food System* by M. Heller and G. Keolian (2000) and *Toward Sustainable Agricultural Systems in the 21st Century* by the National Research Council (2010).

34. National Research Council, A Pivotal Time in U.S. Agriculture, Ch. 2 in *Toward Sustainable Agricultural Systems in the 21st Century*. 2010. The National Academies Press: Washington, DC.
35. K. Hibbard and T. Wilson. Energy, Water and Land Use, Ch. 10 in *Third National Climate Assessment*. 2014. US Global Change Research Program.
36. C. Walthall et al. *Climate Change and Agriculture in the United States: Effects and Adaptation*. USDA Technical Bulletin # 1935. Government Printing Office, Washington, DC.
37. K. Hibbard et al. 2014. Energy, Water, and Land Use. Ch. 10 in *Climate Change Impacts in the United States: The Third National Climate Assessment*, J. M. Melillo, Terese (T. C.) Richmond and G. W. Yohe, eds., U.S. Global Change Research Program, 257–281. doi:10.7930/J0JW8BSF. nca2014.globalchange.gov/report/sectors/energy-water-and-land
38. National Research Council. *Toward Sustainable Agricultural Systems in the 21st Century*. 2010. The National Academies Press, Washington, DC. nap.edu/openbook.php?record_id=12832&page=1
39. T. Lyson. 2004. "Toward a Civic Agriculture" (Ch. 5), in *Civic Agriculture: Reconnecting Farm, Food and Community*. University Press of New England: Lebanon, NH.
40. A. Davis et al. 2012. *Increasing Cropping System Diversity Balances Productivity, Profitability and Environmental Health*. PLoS ONE 7(10): e47149. doi:10.1371/journal.pone.0047149
41. D. Pimentel et al. 2005. "Environmental, Energetic, and Economic Comparisons of Organic and Conventional Farming Systems." *BioScience* Vol. 55, No. 7, pp. 575–82.
42. Rodale Institute. Regenerative Organic Agriculture and Climate Change: A Down-to-Earth Solution to Global Warming. 2014. rodaleinstitute.org/assets/WhitePaper.pdf
43. National Research Council. Adapted from Box 1-7 Contending "Philosophies of Agriculture" (p. 30), in "Understanding Agricultural Sustainability" (Ch. 1), in *Toward Sustainable Agricultural Systems in the 21st Century*. 2010. The National Academies Press, Washington, DC. nap.edu/openbook.php?record_id=12832&page=1

CHAPTER 2

Understanding Exposure

Agriculture is a risky business. Season after season, year after year, successful farmers, growers and ranchers make decisions to reduce the risks to production that emerge as a result of a dynamic web of complex interactions between soils, crops, livestock, pests, weather, finances, regulations and markets. Among the production risks they routinely manage, weather stands out because of the sensitivity of plants and animals to seasonal weather patterns and the variability of those patterns from year to year.

In the last decade or so, producers around the world have reported that weather patterns seem to be changing.[1] They perceive weather becoming more variable and weather extremes increasing in frequency and intensity. These changes in weather patterns have increased weather-related production risks to such an extent that agricultural scientists now recognize a new kind of agricultural risk: *climate risk*. Climate risk is the increased uncertainty caused by more variable patterns of temperature and precipitation and the increase in the frequency and intensity of extreme weather events associated with climate change.[2]

Many, but not all, of the sustainable producers profiled in this book report that changes in weather have increasingly challenged their ability to manage crop and livestock production over the last two decades.[3] In the eastern half of the United States, these farmers report a multitude of changes that have complicated management.

Midwest farmers are used to dealing with extremes in weather, but many report that in the last decade the extremes seem to be getting more extreme and more frequent. Excessive rainfall has complicated fieldwork and caused catastrophic flooding. Weed populations are increasing as the growing season lengthens. More extreme fluctuations in weather and increasingly strong winds are creating challenges for livestock producers. Increased variability in late winter and spring temperatures has complicated fruit production and caused record-breaking tree-fruit crop failures in the region in 2002 and again in 2012.[4] Farmers in the *Northeast* have been challenged since about 2000 by increases in very heavy rainfall, more extreme weather events and catastrophic flooding, along with unusual periods of drought, high temperatures and heat waves. Livestock producers are changing production practices to manage an increase in variability of spring weather, more extreme fluctuations in temperature and a greater number of more extreme weather events. Throughout the *Southeast*, producers are being challenged by increased weather variability and more extreme summer drought and heat waves.

In the western regions of the country, the twenty-first century brought novel weather challenges, mostly associated with drought. Farmers and ranchers in the *Great Plains* report increased variability and extreme temperature fluctuations, warmer summers and more dry periods and drought. Farmers and ranchers in the *Southwest* region have varied perceptions of climate risk— some believe the current fifteen-year drought is unprecedented, while others view it as just part of the region's normal weather variability and extremes. Farmers in the *Northwest* are managing drier summers and falls and wetter springs but have enjoyed more moderate weather overall.

Perceptions of changes in weather patterns are complicated by the locale-specific interactions between the production system and climate change effects and are also strongly influenced by cultural, social, personal and political values.[5] Twenty-one of the twenty-five sustainable producers profiled in this book reported that weather patterns have changed since they began farming and that some of these changes seem to fall outside their experience of normal weather variability. The latest climate change science supports these perceptions.

Climate Change in the Twentieth-Century United States

The US Global Change Research Program (USGCRP) released the Third National Climate Assessment (NCA) in 2014.[6] The assessment, the third in a series of reports to the nation on climate change since 1990, provides a detailed regional analysis of observed climate changes and climate change impacts in the United States over the last century and scenario-based projections through the end of this century. Observed changes over the last century include increasing average temperatures, increasing weather variability, warmer nights and winters, a lengthening of the growing season and an increase in the frequency and intensity of extreme weather events.

Although these trends hold true for most of the United States, the degree of change is quite different depending on location. Climate change has not and will not be experienced the same way throughout the United States, because regional topography interacts with the global climate system to create regional patterns of climate change effects. Without substantial reductions in global emissions of heat-trapping gases, these observed changes in regional climate are projected to increase in pace and intensity through the end of this century.

2.1. The Third National Climate Assessment Is Regionally Based and Examines Thirteen Sectors of the US Economy[7]

The Third National Climate Assessment is a regionally based analysis that organizes the United States and its territories into ten regions. The continental United States is divided into six regions, plus Alaska, Hawaii and the Pacific Islands, the coasts and the oceans, as shown in Figure 2.1. From the

Pacific Northwest to the Shenandoah Valley, the Great Lakes to the Gulf of Mexico, our country's landscapes and communities vary dramatically. One common challenge facing every US region is a new and dynamic set of realities resulting from our changing climate.

Higher temperatures, rising sea levels and more extreme precipitation events are altering the work of farmers, insurance agents, water planners, first responders and many others, influencing economic sectors from coast to coast. Agriculture, energy, transportation and more are all affected by climate change in concrete ways. American communities are contending with these changes now and will be doing so increasingly in the future.

Economic sectors do not exist in isolation but are linked in complex ways. Forest management activities, for example, affect and are affected by water supply, changing ecosystems, impacts to biological diversity and energy availability. Water supply and energy use are completely intertwined, since water is used to generate energy and energy is required to pump, treat and deliver water—which means that irrigation-dependent farmers and urban dwellers are linked as well. Human health is affected by water supply, agricultural practices, transportation systems, energy availability and land use, among other factors—touching the lives of patients, nurses, county health administrators and many others. Human communities are directly affected by extreme weather events and changes in natural resources such as water availability and quality; they are also affected both directly and indirectly by ecosystem health.

The Third National Climate Assessment explores some of these issues within individual sectors but takes a cross-sector approach on others. Seven single-sector chapters focus on:

- Water resources
- Energy production and use
- Transportation
- Agriculture
- Forests
- Human health
- Ecosystems and biodiversity

Six crosscutting chapters address how climate change interacts with multiple sectors. These cover the following topics:

- Energy, water and land use
- Urban infrastructure and vulnerability

FIGURE 2.1. The Nine Regions of the National Climate Assessment. Credit: USGCRP

- Indigenous peoples, lands and resources
- Land use and land cover
- Rural communities
- Biogeochemical cycles

A common theme is that these various sectors are interconnected in many ways. Decisions made in one sector can create a cascade of events that affect resilience to climate change across multiple sectors. This systems approach helps to reveal, for example, how adaptation and mitigation strategies are part of dynamic and interrelated systems. In this way, for example, adaptation plans for future coastal infrastructure are connected with the kinds of mitigation strategies that are—or are not—put into place today, since the amount of future sea level rise will differ according to various societal decisions about current and future emissions of heat-trapping gases.

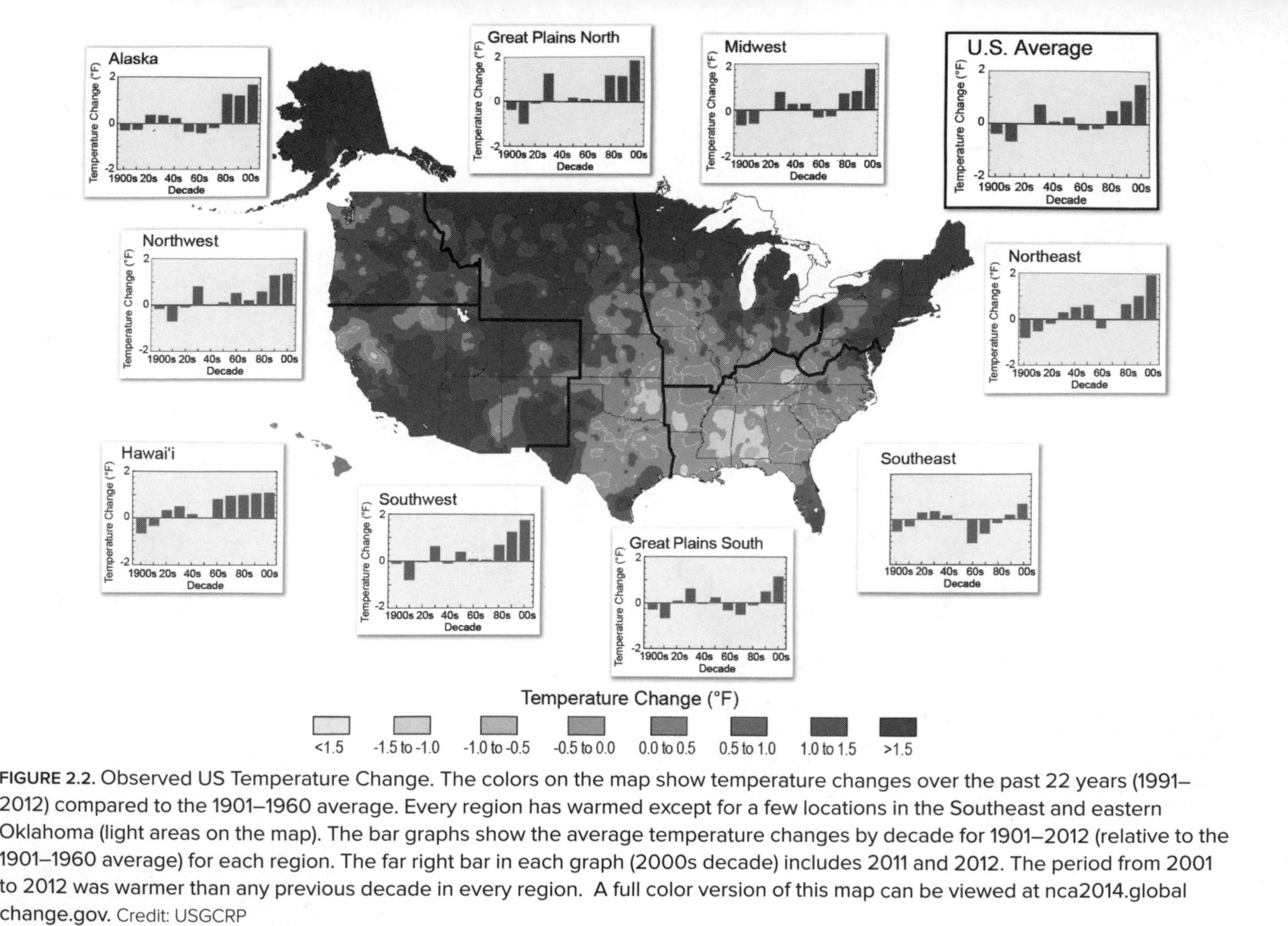

FIGURE 2.2. Observed US Temperature Change. The colors on the map show temperature changes over the past 22 years (1991–2012) compared to the 1901–1960 average. Every region has warmed except for a few locations in the Southeast and eastern Oklahoma (light areas on the map). The bar graphs show the average temperature changes by decade for 1901–2012 (relative to the 1901–1960 average) for each region. The far right bar in each graph (2000s decade) includes 2011 and 2012. The period from 2001 to 2012 was warmer than any previous decade in every region. A full color version of this map can be viewed at nca2014.globalchange.gov. Credit: USGCRP

Changing Temperature Patterns

The climate record shows that average temperatures in the United States have increased over the last century. Figure 2.2 presents regional patterns of temperature change across the United States during the twentieth century as average and decadal changes. More warming has occurred in the western and northern regions of the country than in the Southeast; however, starting about 1980, all regions of the country began warming, and the first decade of the twenty-first century was warmer than any previous decade in every region (see bar graphs in Figure 2.2).

These warming trends are apparent in climate data more relevant to agricultural production. The growing season lengthened during the twentieth century, increasing by about nineteen days in the West, ten days in the North and six days in the Southeast (see Figure 2.3). In

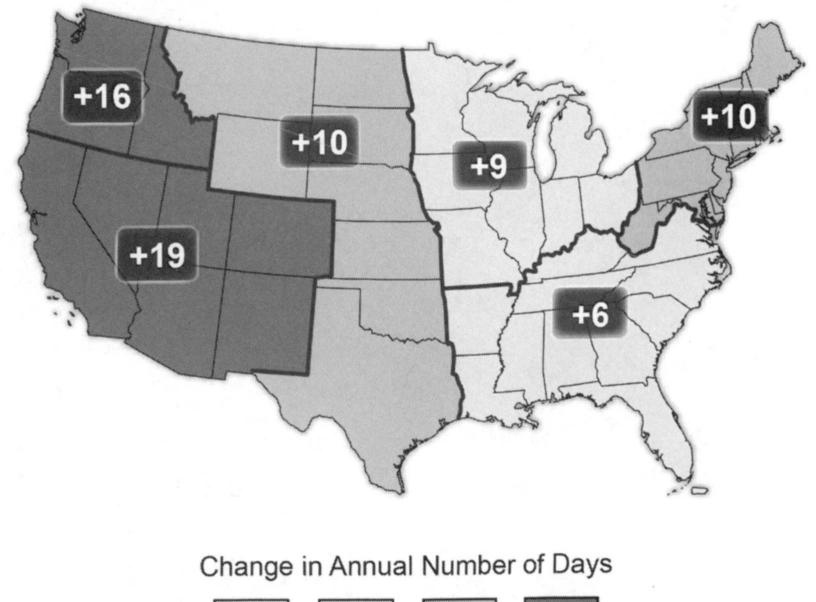

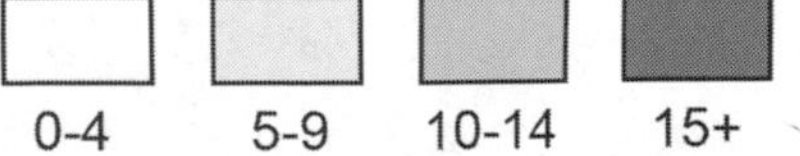

FIGURE 2.3. Observed Changes in the Frost-free Season. The frost-free season length is the period between the last occurrence of 32°F in the spring and the first occurrence of 32°F in the fall. Increases in frost-free season length correspond to similar increases in growing season length. Credit: USGCRP

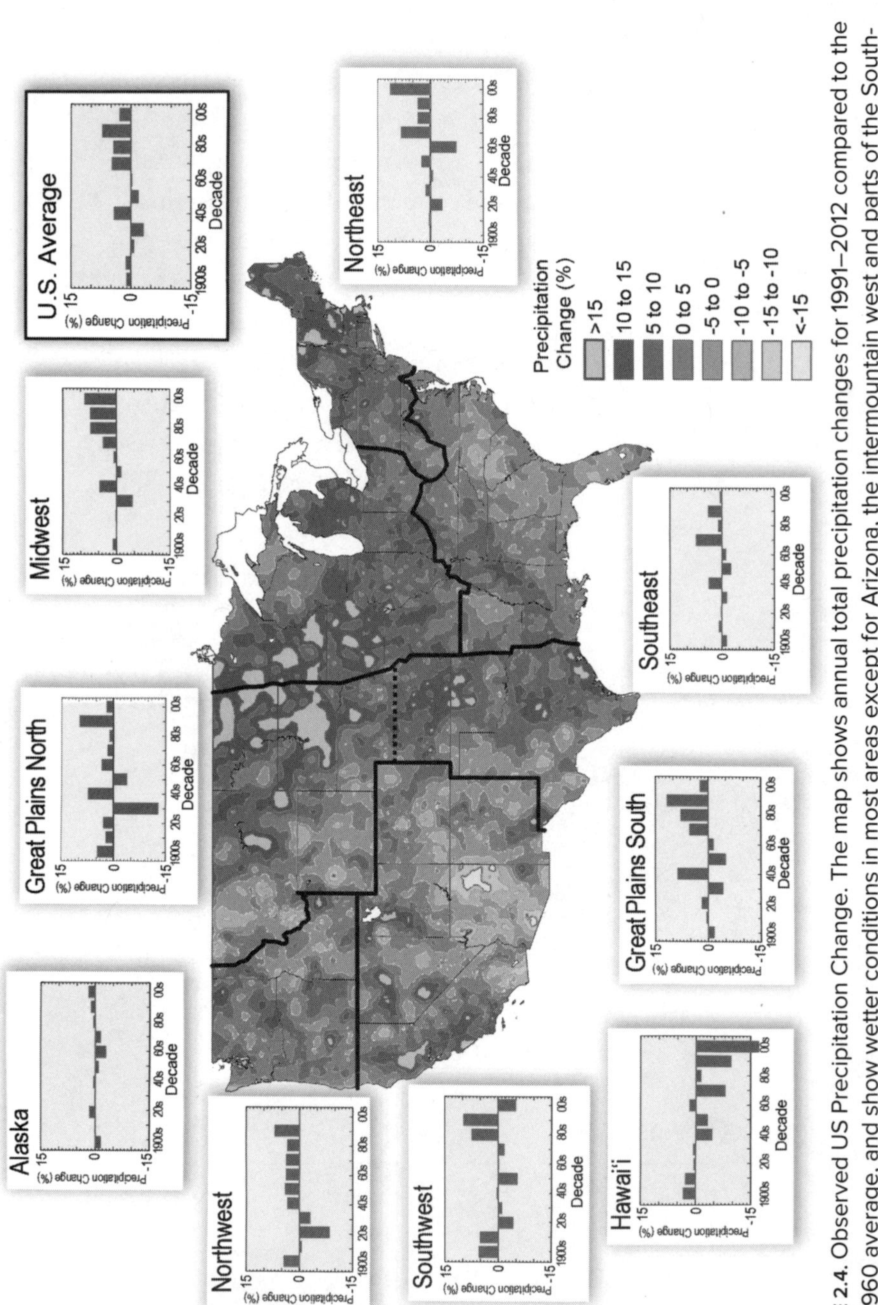

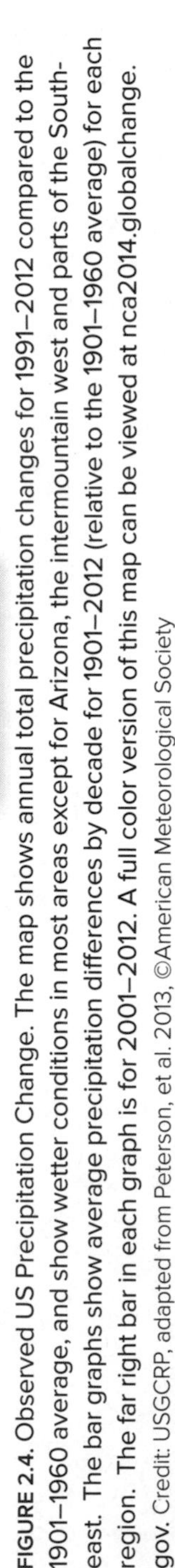
FIGURE 2.4. Observed US Precipitation Change. The map shows annual total precipitation changes for 1991–2012 compared to the 1901–1960 average, and show wetter conditions in most areas except for Arizona, the intermountain west and parts of the Southeast. The bar graphs show average precipitation differences by decade for 1901–2012 (relative to the 1901–1960 average) for each region. The far right bar in each graph is for 2001–2012. A full color version of this map can be viewed at nca2014.globalchange.gov. Credit: USGCRP, adapted from Peterson, et al. 2013, ©American Meteorological Society

addition, minimum temperatures are increasing faster than maximum temperatures, leading to more rapid warming at night and in winter.[8] These kinds of changes in daily and seasonal temperature patterns have the potential to disrupt agricultural production because of the sensitivity of crops, livestock and pests to temperature.[9]

Changing Precipitation Patterns

Average precipitation increased during the twentieth century, although this increase is partly the result of two major droughts earlier in the century (in 1930 and 1960). Figure 2.4 illustrates the changes in average precipitation across the United States through the twentieth century. Like the temperature change map (Figure 2.2), Figure 2.4 presents observed change in average precipitation over the twentieth century (map) and decadal precipitation (bar charts).

While average US precipitation was higher than the baseline over the last forty years, the patterns of change in average precipitation are not as clear as those for temperature. Significant regional patterns of change are more clearly illustrated by observed changes in very heavy precipitation and flood magnitude trends (Figures 2.5 and 2.6). These changes in the frequency and intensity of precipitation events support the perceptions of many US farmers that more frequent and intense rainfall and associated flooding have increased climate risk and complicated management of crops and livestock, particularly in the Midwest and the Northeast.

Projected Climate Changes Through the 21st Century

Climate risk is expected to become an increasingly important factor in agricultural production in the years ahead as the pace and intensity of climate change accelerates through this century.[12] Over the next few decades, the entire United States will continue to warm by about 2 to 4 degrees Fahrenheit. This rate of warming represents an increased rate of change over that observed in the last few decades and is a substantially greater rate of change than was experienced over the twentieth century.

Projection of climate changes beyond mid-century are less certain, because the rate of change is driven largely by the level of heat-trapping gases being released to the atmosphere by human activities. The NCA

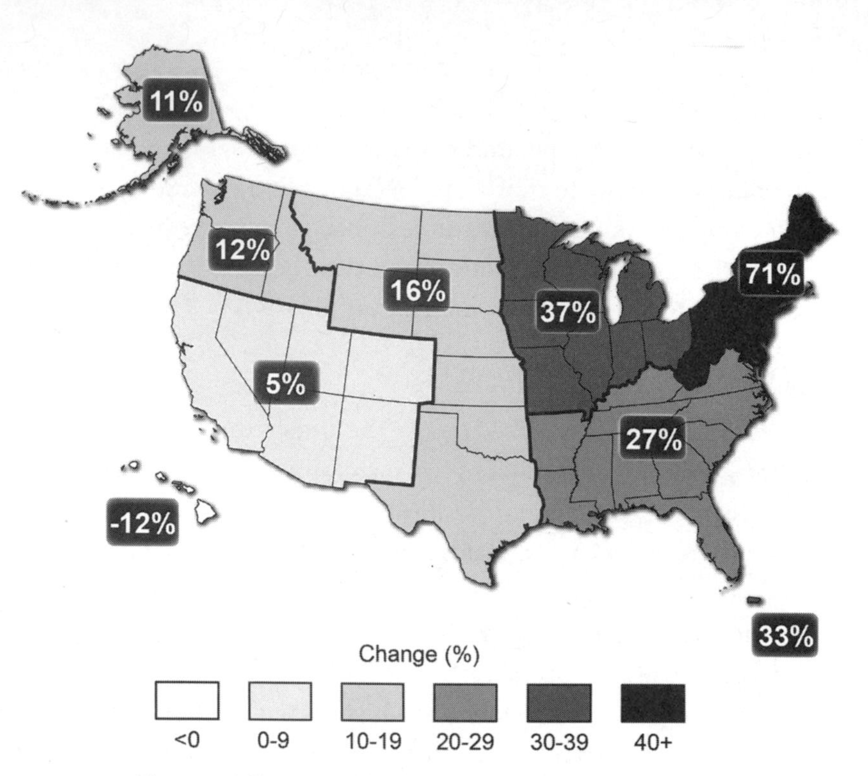

FIGURE 2.5. Observed Change in Very Heavy Precipitation. The map shows percent increases in the amount of precipitation falling in very heavy events, defined as the heaviest 1% of all daily events, from 1958 to 2012 for each region of the continental US. These trends are larger than natural variations for the Northeast, Midwest, Southeast, Puerto Rico, the Great Plains and Alaska. Credit: USGCRP

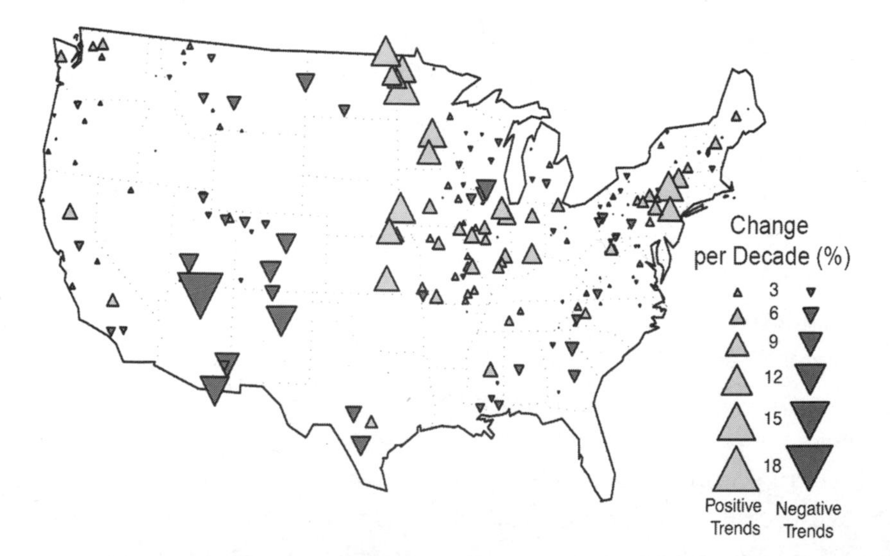

FIGURE 2.6. Trends in Flood Magnitude. Trend magnitude (triangle size) and direction (pointing up = increasing trend, pointing down = decreasing trend) of annual flood magnitude from the 1920's through 2008. Credit: USGCRP, adapted from Peterson, et. al. 2013, ©American Meteorological Society.

reports on four possible scenarios based on emissions levels, ranging from a rapid reduction in emissions—a 70 percent reduction by 2050 and more reductions through the end of the century—to continued increases in emissions through the end of the century.

Projections for the latter part of the twenty-first century suggest that average temperatures will rise about 4 degrees Fahrenheit if substantial reductions in emissions are achieved and by as much as 10 degrees Fahrenheit if current trends in emissions continue.[13] Temperatures are projected to increase on both the coldest and warmest days of the year throughout the nation. Heat waves are projected to increase throughout most of the United States, and droughts are likely to become more intense in the Southwest. The growing season will continue to lengthen, increasing by as much as a month in many parts of the nation and as much as two months in the West, while the number of frost days will decline by twenty to thirty days in most of the nation and by even more in the West (see Figure 2.7). Dry periods will lengthen, with the greatest increases expected in the Northwest, Southwest and southern Great Plains, and hot nights are expected to increase by more than eighty per year across the southern US by the end of the century (see Figure 2.8).

There will likely be more winter and spring precipitation in the northern parts of the US and less precipitation in the Southwest, while summer and fall precipitation is likely to remain about the same or decrease in most regions. Both the frequency and intensity of heavy rainfall events are projected to increase.

Based on these projections, many of the weather-related challenges reported by the sustainable producers featured in this book across the United States are likely to continue and are expected to intensify through mid-century. The pace of change is also projected to increase, adding additional uncertainty to farm and ranch management. The rapid increases in the pace and intensity of climate change represents a novel risk management challenge to US agriculture. New strategies will be needed to sustain agricultural production if climate change intensifies as projected through the twenty-first century.[14]

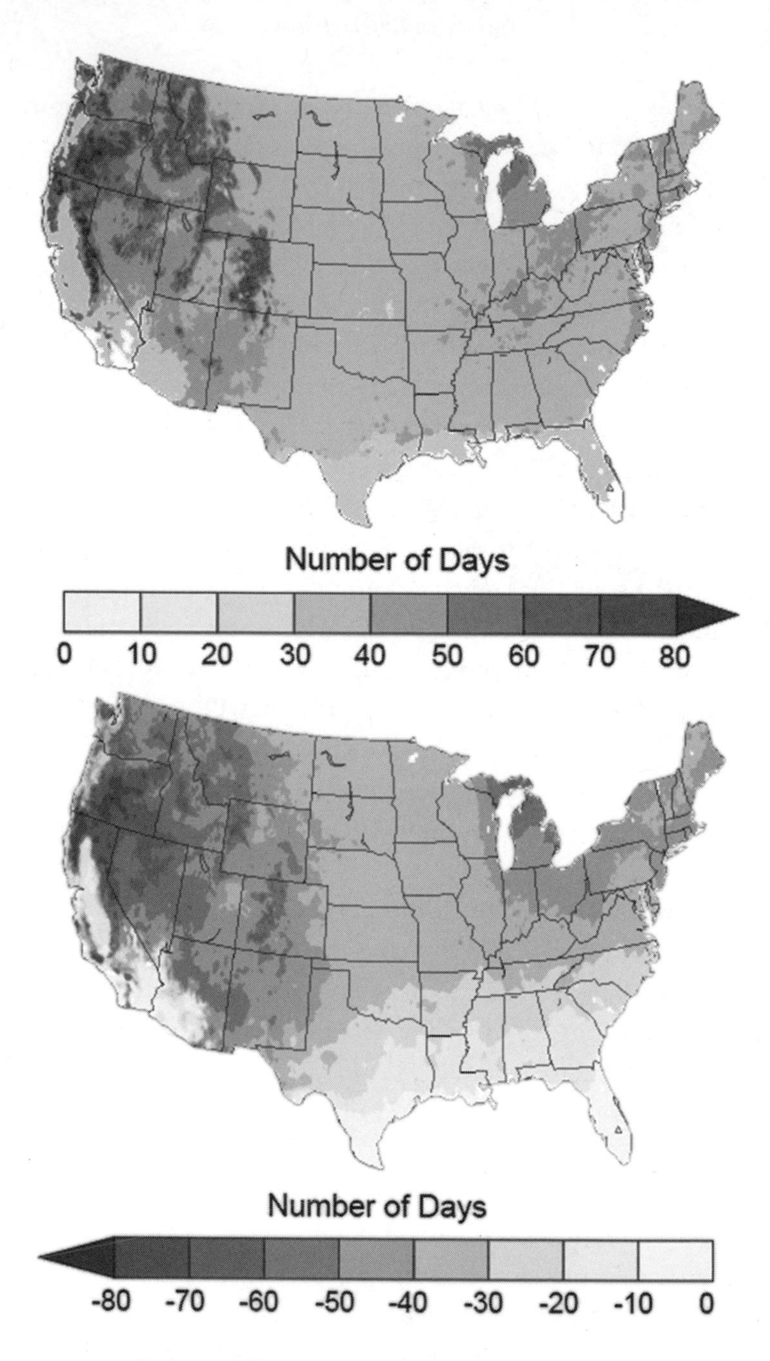

FIGURE 2.7. Projected Changes in Frost-free Season and Number of Frost Days. These maps show projected changes in the frost-free season and number of frost days at the end of the century (2070–2099) compared to 1971–2000, assuming continued increases in heat-trapping gases through mid-century. A few areas in southern California and south Florida are projected to have no freezes during the period 2070–2099 (white areas). Credit: USGCRP

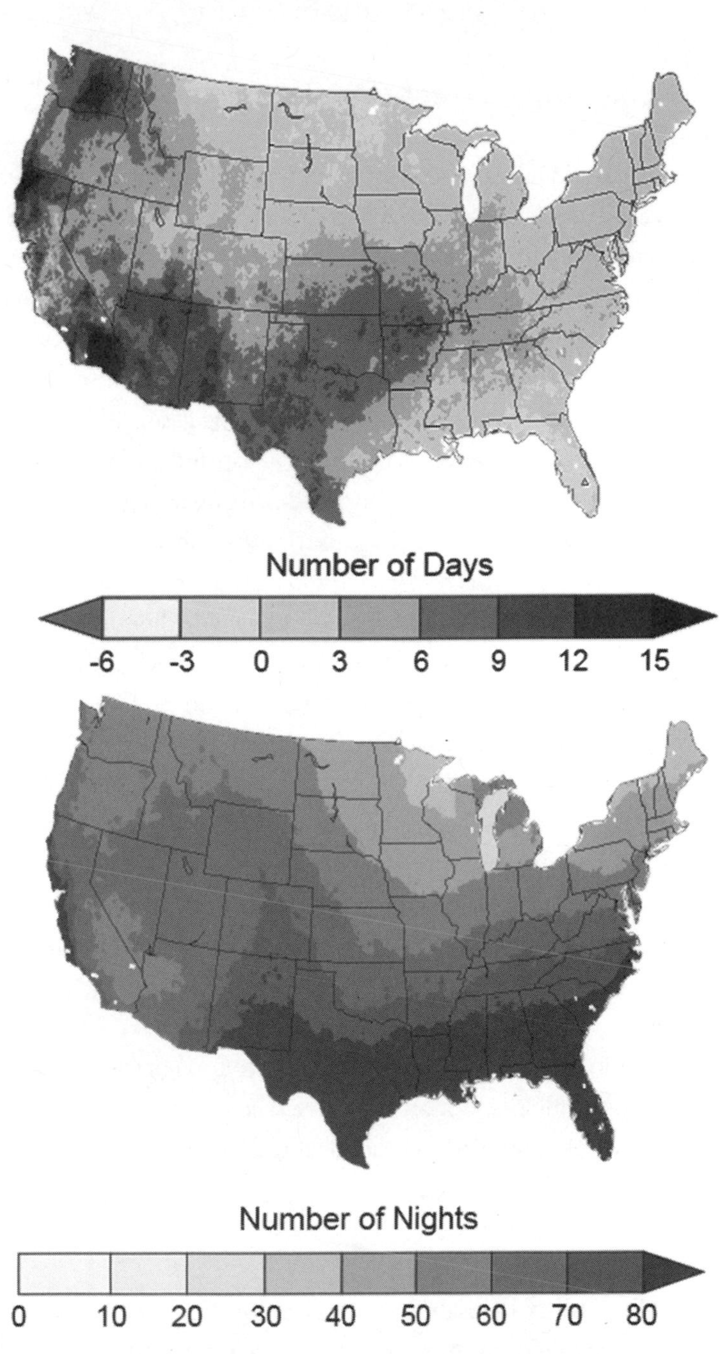

FIGURE 2.8. Projected Changes in Dry Periods and Hot Nights. These maps show projected changes in dry periods (consecutive days with less than 0.01 inches of precipitation) and hot nights (as nights with a minimum temperature higher than 98% of the minimum temperatures between 1971 and 2000) at the end of the century (2070–2099) compared to 1971–2000, assuming continued increases in heat-trapping gases through mid-century. Credit: USGCRP

2.2. Heat-trapping Gases Are Warming the Earth[15]

Many natural and human factors influence global temperatures. These factors are called *drivers* because they can cause the Earth's temperature to change direction, vary from the current path, rise, drop or stay the same. These temperature changes influence global climate. Natural global climate drivers include the energy from the Sun; the addition of tiny particles in the atmosphere from volcanic eruptions, wind-blown dust and salt spray; and natural processes that release or absorb heat-trapping gases, like the growth and death of plants. Human global climate drivers include burning coal, gas and oil in power plants and cars; cutting down and burning forests; the release of soil, soot and other particles into the atmosphere; and changes in land use that cause the Earth to absorb or reflect more of the Sun's heat.

Some of these factors have a warming effect, while others lead to cooling. The effect of each of these factors on global temperature can be directly compared when expressed as "forcing" in units of watts per square meter. Figure 2.9 shows that when the most important natural and human-induced factors are accounted for, the human-released heat-trapping gases in the atmosphere appear to be the main factor causing the global warming observed over the last half of the twentieth century. These gases include *carbon dioxide* released by our fossil-fueled lifestyle; *methane* released by industrial agriculture, mining and landfills; and *nitrous oxide* released by the use of agricultural fertilizers and the burning of fossil fuels.

The conclusion that human activities appear to be an important factor in recent warming is supported by observed changes in global temperature over the twentieth century (Figure 2.10).

Managing Climate Risk

Climate vulnerability describes the degree to which a system (for example, a farm, rural community or nation) is susceptible to the adverse effects of climate. Vulnerability has three component parts: exposure, sensitivity and adaptive capacity (see Figure 2.11). *Exposure* describes the type and intensity of climate change effects likely to be experienced

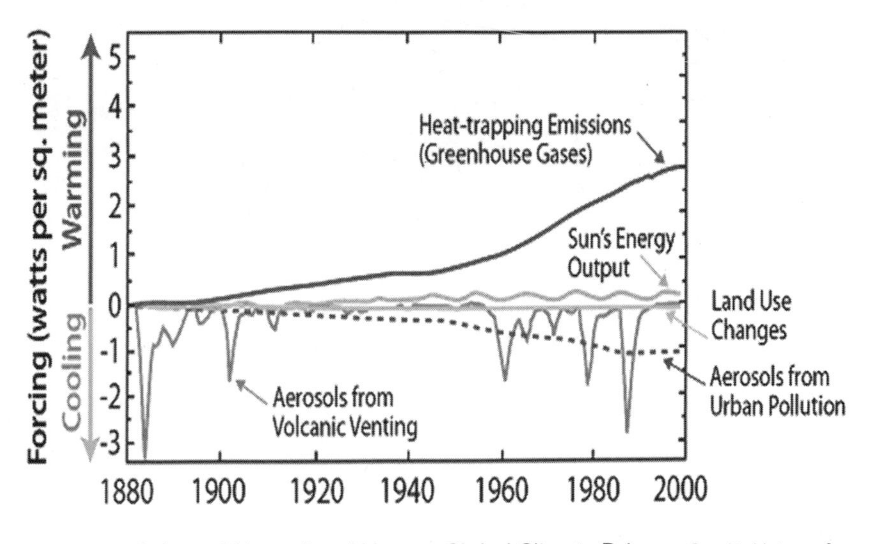

FIGURE 2.9. Selected Natural and Human Global Climate Drivers. Credit: Union of Concerned Scientists

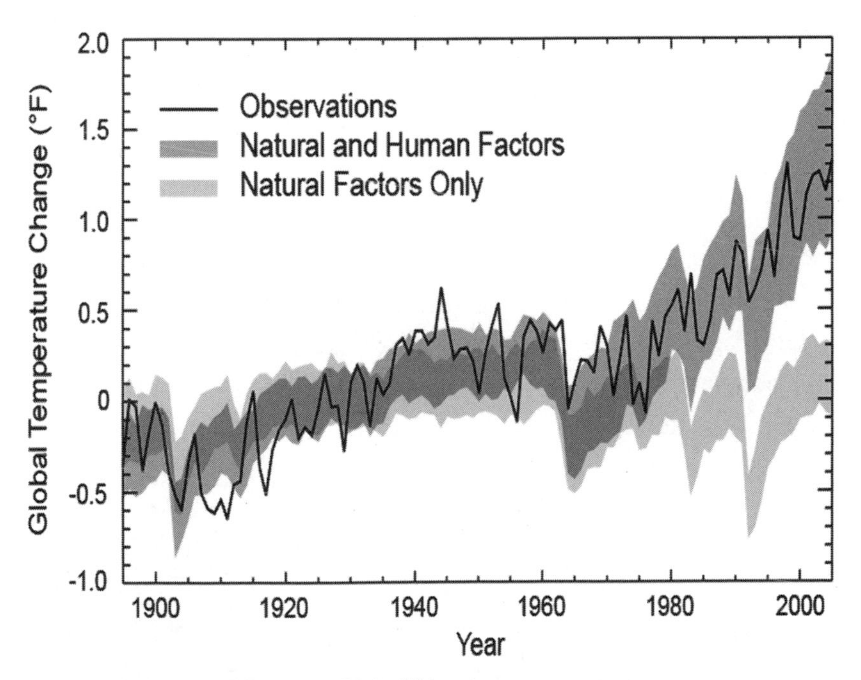

FIGURE 2.10. Human Effects on Global Warming. Credit: USGCRP

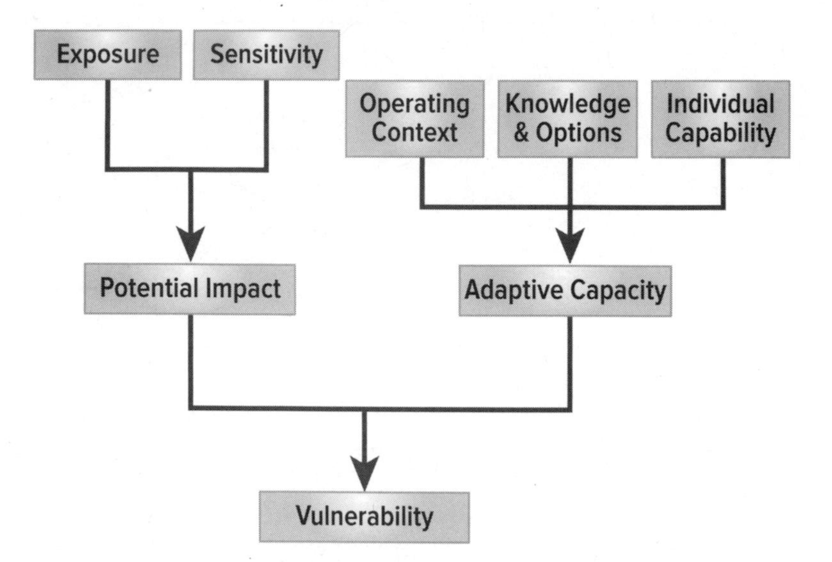

FIGURE 2.11. Linked Human and Biophysical Factors that Determine the Vulnerability of Agricultural Systems to Climate Change. The potential impact of climate change on a farm or ranch is determined by the interaction between its exposure and sensitivity to climate change effects. Exposure describes the type and intensity of climate change effects likely to be experienced at a specific location. Sensitivity describes the degree to which elements of the farm or ranch system respond, negatively or positively, to climate-related events. Adaptive capacity describes the potential of the farm or ranch system to buffer climate change effects. Credit: USDA

at a specific location. Think of this as the local experience of climate change—the interaction between the global climate system and the regional landscape at a specific location. For example, a farm located in a floodplain will be more likely to experience flooding during heavy rainfall events than a farm located outside of a floodplain. *Sensitivity* expresses the degree to which elements of the farm system respond, negatively or positively, to climate-related events. For example, fruit trees are very sensitive to chill damage caused by variable spring weather, while early spring vegetable crops like kale are not. *Adaptive capacity* describes the ability of the farm or ranch system to moderate potential damages, take advantage of opportunities or cope with the consequences of climate change. For example, farms that produce a diverse mix of crops tend to enjoy more stable yields over time compared to farms that produce only one or two crops each year.

These three distinct components of vulnerability require different management approaches to reduce climate risk. Together, they determine the vulnerability of a farm or ranch to climate change effects. This way of thinking about climate risk helps to put a focus on the two components of climate vulnerability that can be managed most directly by producers—sensitivity and adaptive capacity—while reminding us of the connections between exposures and the human activities that drive global warming.

Understanding Agricultural Exposure

The NCA reports observed and projected changes in weather at mid-century and at the end of the century for nine regions of the United States. Exploring the maps of observed and projected regional changes in weather patterns and other climate data presented in the NCA helps to identify key climate change exposures for farms and ranches in each region and provides insight into the perceptions of changing weather patterns reported by the producers profiled in this book.

The *Northwest*[16] region has experienced few climate change effects to date. Average temperatures have increased slowly over the last century. Average precipitation has increased, particularly in the spring, and heavy downpours have increased somewhat. Observed regional warming has reduced summer flow in river basins fed by snowmelt.

Average temperatures in the Northwest region are expected to continue to rise throughout this century, and projected changes in precipitation will vary from slight decreases to large increases in winter, spring and fall. By mid-century, summer flows are projected to be substantially lower in many Northwest rivers. Summer precipitation is projected to decrease through the century, while an increase in extreme precipitation events will cause more flooding in transient and rain-fed river basins. Although the overall consequences of climate change will probably be lower in the Northwest than in other regions of the United States, the sustainability of some Northwest agricultural sectors will be challenged by projected increases in average temperatures, hot weather episodes and uncertainty in summer water supply.

The *Southwest*[17] region has recently experienced some of the most extreme climate change impacts in the nation. The average temperature over the second half of the twentieth century was the warmest in six hundred years, and the decade from 2001 to 2010 was the warmest in the last century, with more heat waves and fewer cold snaps. Unlike temperatures, changes in precipitation over the last fifty years were more variable across the region, with only small changes in average precipitation in most areas; however, across most of the Southwest, less late winter snow, earlier snowmelt and earlier arrival of most of the year's streamflow have reduced river flows. Crop and livestock productivity have been boosted in some locations by the lengthening growing season in combination with less frequent cold snaps, but more frequent heat waves and declining water resources have presented challenges to agricultural production in most of the Southwest region. All of these changes are projected to continue and intensify through mid-century.

Regional annual temperatures in the Southwest are projected to continue to rise through this century, with the greatest increases in the summer and fall. Summertime heat waves will become longer and hotter, and the recent trend of decreasing wintertime cold snaps is projected to continue. Reductions in precipitation are consistently projected for the southern part of the region, as is a reduction in spring precipitation, while projected precipitation in the northern part of the region and the other seasons in the South are mixed. Late winter and early spring snowpack are projected to decrease through this century, reducing runoff to surface waters and soil moisture content, while an increase in winter flood risk is projected for the region. Projected reductions in winter chill periods and an increase in hot summer conditions may reduce the productivity of crops and livestock.

The *Great Plains*[18] is a diverse region where dramatic weather creates multiple climate and weather hazards. Climate variability already stresses Great Plains agriculture, and current trends and projections through mid-century suggest that more intense droughts, heavy rainfalls and heat waves will become more frequent. Great Plains residents already contend with weather challenges from winter storms, extreme heat and cold, severe thunderstorms, drought and flood-producing rainfall. Summers are long and hot in the south; winters are long and often severe in the north. The average annual temperature increase in North Dakota has been the fastest in the contiguous United States and is driven mainly by warming winters. Across much of the Great Plains region, annual water loss from transpiration by plants and evaporation is higher than annual precipitation, making these areas particularly susceptible to drought.

Projections suggest that the region will grow warmer and the growing season will lengthen by an average of twenty-four days by mid-century. This warming will result in significant increases in summer days over 100 degrees Fahrenheit and winter days over 60 degrees throughout the region. Precipitation is projected to increase in the north and decrease in the south through the end of this century, and the number of days with heavy precipitation (at least one inch) is expected to increase by mid-century, especially in the north. Days with little or no rainfall will be less common in the north. By contrast, large parts of Texas and Oklahoma are projected to see more days with no precipitation through mid-century. Increased snowfall, rapid spring warming and intense rainfall can combine to produce devastating floods, as is already common along the Red River of the North. Increased drought frequency and intensity may turn marginal lands into deserts.

Agriculture dominates land use in the *Midwest*,[19] with more than two-thirds of the region's land devoted to crop production. A lengthening growing season, warmer winters, increased incidence of late spring freezes, increased heat and drought in summer and more frequent heavy rainfall and catastrophic flooding are already complicating agricultural production in the region. Tree fruit producers sustained major crop losses after late spring freezes in 2002 and 2012, and high nighttime temperatures reduced corn yields in 2010 and 2012.[20] These climate change effects are projected to increase in intensity through this century.

Over the next 30 years, future crop yields in the region are most likely to be negatively impacted by extreme weather events that damage crops directly or by interfering with a critical development phase. As a result, projected increases in yields associated with increased temperatures, longer growing seasons or elevated atmospheric carbon dioxide concentrations could be offset by more frequent damage from late spring freeze events, late planting because of wet soil conditions, the disruption of critical crop developmental phases and reductions in crop quality. Extremes in precipitation are projected to intensify across all seasons, with the likelihood of both increasing heavy rain and snow events and more droughts. Increased runoff and flooding will reduce surface water quality because of increased soil erosion and sediment loading to surface waters.

The *Northeast*[21] has a diverse climate that varies with latitude, proximity to the coast, elevation and season. Warm humid summers give way to cold frigid winters, and extreme events such as ice storms, floods, droughts, heat waves, major storms and hurricanes are common. Over the last century, annual temperatures in the region have increased by about two degrees Fahrenheit, annual precipitation has increased by about five inches, and sea level rise of about one foot has increased damages from coastal flooding. The Northeast has experienced a greater increase in extreme precipitation than any other region in the United States.

Agriculture in the region is already stressed by climate change. More frequent and intense extreme precipitation events damage crops directly and reduce crop yields by interfering with timely field operations like planting, cultivating, pest management and harvest. Warmer winters have increased the risk of frost and freeze damage in perennial crops, causing major crop losses to damaging frosts in 2007 and 2012. Longer growing seasons have increased weed and pest pressures in the region.

Over the next thirty years, more high temperatures and longer heat waves, warmer and more variable winter and spring weather, more dry periods and drought, and more frequent heavy rains and damaging storms will create increasingly stressful conditions for agricultural production. Temperatures will increase, with the southern part of the region experiencing as much as two more months of temperatures above ninety-five degrees Farenheit. Winter precipitation is expected to increase, most strongly in the northern part of the region. Seasonal drought risk is expected to increase in summer and fall as higher temperatures lead to greater evaporation and earlier winter and spring snowmelt. By the end of the century, projected sea level rise will more than triple the risk of dangerous coastal flooding throughout the region.

The *Southeast*[22] is particularly vulnerable to hurricanes, which have increased in intensity since the 1970s and explain, in part, why the Southeast has suffered more billion-dollar disasters than any other region in the United States over the last thirty years. Although average annual temperatures have cycled between warm and cool periods over the last century, this region has experienced the most rapid warming in the nation since the 1970s. Much of this warming has occurred in the summer months, increasing the number of days above 95 degrees Fahrenheit and the number and intensity of extreme summer heat events.

These changes in weather in the *Southeast* are projected to increase in intensity through mid-century, with the greatest warming projected for the southern and western parts of the region. As sea levels rise through the century, saltwater intrusion into freshwater supplies will limit crop production in some areas, particularly in Florida and southern Louisiana. Increasingly long, hot summers will likely cause a decline in crop and livestock production in the region as a result of more intense heat stress accompanied by longer and more intense drought periods.

Reducing Exposure

The only option for reducing exposure is to take action to slow or reverse—to mitigate—global warming. In a vulnerability context, mitigation has a very specific meaning, referring to any activity that leads to a reduction of heat-trapping gases in the atmosphere. This reduction can be accomplished by releasing less heat-trapping gases (reducing emissions) or removing heat-trapping gases from the atmosphere (sequestration).

Mitigation is a crucial component of broad efforts to manage climate risk, because mitigation helps to slow the pace of climate change, which can reduce the cost and complexity of adaptation to local climate exposures. Mitigation is most successful coordinated through international collaboration, because investments in mitigation have global benefits shared by all people, regardless of their investment in the mitigation effort.[23] Adaptation is most successful coordinated at the local level, because effective adaptive actions address the unique place-based impacts of climate change arising from the specific sensitivities of a system to specific climate exposures.

The most cost-effective adaptive actions are sometimes those that also serve to mitigate climate change and so reduce climate exposures. These "win-win" and "low- or no-regrets" adaptation options are often low cost or provide multiple benefits.[24] Agriculture offers many opportunities for these kinds of adaptations because of the enormous sequestration potential of high-quality soils[25] and climate protection potential of agricultural landscapes. High-quality soils buffer variability and extremes in precipitation while sequestering heat-trapping gasses. The restoration of wetlands and riparian areas in agricultural lands and the use of perennial plantings both sequester heat-trapping gases and provide climate protection services through reduced inland flooding and moderation of extremes of temperature and wind. Slowing global warming through mitigation is critically important to the success of climate change adaptation efforts. A reduction in the pace and intensity of climate change will reduce the costs of adaptation and provide more time to develop effective adaptation strategies.[26]

Endnotes

1. For example, in Canada, C. Bryant et al., "Adaptation in Canadian Agriculture to Climatic Variability and Change." 2000. *Climatic Change* 45:181; in Europe, J. Olesen et al., "Impacts and Adaptation of European Crop Production Systems to Climate Change." 2011. *Europ. J. Agronomy* 34:96–112; in Australia, N. Marshall. "Understanding Social Resilience to Climate Variability in Primary Enterprises and Industries." 2010. Global Environmental Change 20:36–43; in New Zealand, G. Kenny. "Adaptation in Agriculture: Lessons for Resilience from Eastern Regions of New Zealand." 2011. Climatic Change 106:441–462; in the US, J. Arbuckle et al. "Farmer Beliefs and Concerns about Climate Change and Attitudes Toward Adaptation and Mitigation: Evidence from Iowa." 2013. Climatic Change. DOI 10.1007/s10584-013-0700-0; in the US, R. Rejesus et al. "Agricultural Producer Perceptions of Climate Change." 2013. U.S. J. of Ag. Applied Econ. 45:701–718.
2. M. Howden et al. Adapting Agriculture to Climate Change. 2007. *Proc. of the National Acad. of Sci.* 104 (50): 19691.
3. The twenty-five producers featured in this book were interviewed by phone between August 2013 and March 2014. They were asked to describe their perceptions of changes in weather over the time that they have managed their current farm or ranch. If they reported changes, they were asked to describe

how those changes complicated their farm management and how they have adapted to them.

4. N. Rothwell, M. Woods and P. Korson. 2013. "Assessing and Communicating Risks from Climate Variability for the Michigan Tart Cherry Industry." In: *Project Reports*. D. Brown, D. Bidwell and L. Briley, eds. Great Lakes Integrated Sciences and Assessments Center.
5. C. Walthall et al. Adapting to Climate Change, Ch. 6, in *Climate Change and Agriculture in the United States: Effects and Adaptation*. 2012. USDA Technical Report 1935.
6. U.S. Global Change Research Program. The Third National Climate Assessment. 2014. nca2014.globalchange.gov/
7. Adapted from "Introduction to Regions" (p. 369) and "Introduction to Sectors" (p. 68), *Third National Climate Assessment*. 2014. USGCRP. nca2014.globalchange.gov/downloads
8. See High and Low Temperatures, in *Climate Change Indicators in the United States*, Environmental Protection Agency. 2014. epa.gov/climatechange/science/indicators/weather-climate/high-low-temps.html
9. C. Walthall et al. 2012. Climate Change Science and Agriculture, Ch. 4, in *Climate Change and Agriculture in the United States: Effects and Adaptation*. USDA Technical Report 1935.
10. J. Peterson et al. 2013. Monitoring and Understanding Changes in Heatwaves, Coldwaves, Floods and Droughts in the United States: State of Knowledge. *Bulletin of the American Meteorological Society* 94: 821–834.
11. Ibid.
12. J. Hatfield et al. 2014. Ch. 6: Agriculture. *Climate Change Impacts in the United States: The Third National Climate Assessment*, J. Melillo, T. Richmond, and G. Yohe (eds.). U.S. Global Change Research Program, 150–174. doi:10.7930/J02Z13FR.
13. J. Melillo, T. Richmond, G. Yohe, eds. Chapter 2: Our Changing Climate. *Climate Change Impacts in the United States: The Third National Climate Assessment*, U.S. Global Change Research Program, 19–67.
14. J. Hatfield et al. Agriculture, Ch. 6 in *Climate Change Impacts in the United States: The Third National Climate Assessment*, J. Melillo, T. Richmond, and G. Yohe (eds.). 2014. U.S. Global Change Research Program, 150–174. doi:10.7930/J02Z13FR.
15. Adapted from *How Do We Know that Humans Are the Major Cause of Global Warming?* Union of Concerned Scientists, 2009. ucsusa.org/global_warming/science_and_impacts/science/human-contribution-to-gw-faq.html
16. P. Mote et al. Northwest. Ch. 21 in *Climate Change Impacts in the United States: The Third National Climate Assessment*, J. Melillo, T. Richmond

and G. Yohe (eds). 2014. U.S. Global Change Research Program, 487–513. doi:10.7930/J04Q7RWX

17. G. Garfin et al. Southwest. Ch. 20 in *Climate Change Impacts in the United States: The Third National Climate Assessment,* 2010. J. Melillo, T. Richmond, and G. Yohe, eds., U.S. Global Change Research Program, 462–486. doi:10.7930/J08G8HMN.
18. M. Shafer et al. Great Plains. Ch. 19 in *Climate Change Impacts in the United States: The Third National Climate Assessment.* J. Melillo, T. Richmond and G. Yohe (eds). 2014. U.S. Global Change Research Program, 441–461. doi:10.7930/J0D798BC.
19. S. Pryor et al. Midwest. Ch. 18 in *Climate Change Impacts in the United States: The Third National Climate Assessment,* J. Melillo, T. Richmond, and G. Yohe, eds., 2014: U.S. Global Change Research Program, 418–440. doi:10.7930/J0J1012N.
20. J. Hatfield et al. 2014. Agriculture. Ch. 6 in *Climate Change Impacts in the United States: The Third National Climate Assessment,* J. Melillo, T. Richmond, and G. Yohe (eds.). U.S. Global Change Research Program, 150–174. doi:10.7930/J02Z13FR.
21. R. Horton et al. Northeast. Ch. 16 in *Climate Change Impacts in the United States: The Third National Climate Assessment,* J. Melillo, T. Richmond, and G. Yohe, eds., 2014. U.S. Global Change Research Program, 371–395. doi:10.7930/J0SF2T3P.
22. L. Carter et al. Southeast and the Caribbean. Ch. 17 in *Climate Change Impacts in the United States: The Third National Climate Assessment,* J. Melillo, T. Richmond, and G. Yohe, eds. 2014. U.S. Global Change Research Program, 396–417. doi:10.7930/J0NP22CB.
23. C. Walthall et al. Adapting to Climate Change. Ch. 6 (p. 126) in *Climate Change and Agriculture in the United States: Effects and Adaptation.* 2012. USDA Technical Bulletin 1935.
24. A. Brown et al. 2011. Managing Adaptation: Linking Theory and Practice. UK Climate Impacts Programmwe, Oxford, UK.
25. Regenerative Organic Agriculture and Climate Change: A Down-to-Earth Solution to Global Warming. 2014. Rodale Institute. rodaleinstitute.org/assets/WhitePaper.pdf
26. C. Walthall et al. Adapting to Climate Change. Ch. 6 (p. 126) in *Climate Change and Agriculture in the United States: Effects and Adaptation.* 2012. USDA Technical Bulletin 1935.

CHAPTER 3

Understanding Sensitivity

In the context of climate change vulnerability, sensitivity describes the degree to which a system is affected, either positively or negatively, by climate-related effects. In an agricultural production system, these effects may be direct, such as a change in crop growth in response to increased temperatures, or indirect, such as flood damage to crops caused by an increase in heavy rainfall events. Agricultural system sensitivities are associated with the way the production system—the specific crops and livestock, infrastructure, landscape and people—responds to climate change effects.

Understanding the sensitivity of an agricultural production system involves identifying the elements of the farm or ranch that are most responsive to climate effects and exploring potential responses. Some questions that can help guide a sensitivity assessment include:[1]

What kinds of exposures are likely on the farm or ranch? For example, consider landscape relationships. Is any part of the farm in a floodplain or open to prevailing winds?

Is the farm or ranch subject to existing stress? For example, consider farm assets such as natural resource conditions or finances. Are water supplies variable? Is the business struggling financially?

Will exposure push demand for resources above supply? For example, consider equipment needs for timely field operations. Will increased variability in precipitation shorten fieldwork windows to the point that more equipment is needed to complete the work in a timely manner? Is there enough forage production to meet needs during extended dry period or drought?

Are the plant and animal species under management near the limits of their productive resource space? For example, think about the performance of existing species on the farm or ranch. Have changes in temperature or precipitation caused the production of crops or livestock to become more difficult or costly? Have harvested yield and quality of crops and livestock become more variable or declined?

What effect thresholds[2] or tipping points are associated with the production system? These are events or conditions that dramatically change resource quality or production system performance. For example, think about temperature and precipitation limits for an existing species on the farm or ranch. What temperature or water supply conditions would result in complete loss of product? How about a long-term asset such as a cowherd or an orchard? How many years could the farm or ranch continue without any income?

What factors limit the ability of the production system to adapt to climate effects? Limits can be financial, social, cultural or natural resource-based. For example, a farm managing more limited water supplies may not be able to adapt if financial resources are not available to invest in more efficient irrigation or existing water supplies fall below levels needed for crop production.

An assessment of crop and livestock sensitivities to climate change effects offers a useful starting point for adaptation planning because quite a lot is known about the resource conditions that promote optimum growth and development.

Crop Sensitivities

Direct Effects: Atmospheric Carbon Dioxide, Temperature and Water

Climate change influences crop yield and quality through the three basic plant processes that determine plant growth: photosynthesis, respiration and transpiration. All three processes are highly responsive to atmospheric carbon dioxide levels, seasonal temperature fluctuations and maximums and minimums, and the effects of precipitation on soil moisture, plant-available water and relative humidity.

Carbon dioxide (CO_2) levels in the global atmosphere do not vary a great deal from location to location, but temperature and precipitation are highly variable across a wide range of temporal and spatial scales. At the farm or ranch scale, CO_2 levels will be close to global average trends, while local temperature and precipitation vary quite considerably from regional averages. Both temperature and precipitation interact with local topography and landscape features such as windbreaks to create complex and dynamic microclimates.

Plants require atmospheric CO_2 and water to carry out photosynthesis, the process they have evolved to create the energy and basic materials they need for their growth and development. In this process, plants capture and store sunlight energy in simple sugars made with water taken up from the soil and CO_2 absorbed from the atmosphere. Plants use the process of cellular respiration to release this stored energy to fuel many kinds of physiological processes. The balance between photosynthesis and respiration determine a plant's rate of growth and potential productivity. These processes are both profoundly influenced by temperature and water availability.

Photosynthesis and Respiration

Plants gather the three raw materials needed for photosynthesis from their immediate environment. They take up energy from sunlight falling on their leaves, water in the soil through their roots and carbon dioxide from the atmosphere through small openings in their leaves called stomata.

The simple sugars that plants create through photosynthesis serve two key roles in their growth and development. These simple sugars are the building blocks for more complicated carbon compounds that form structural materials such as starch and cellulose, as well as proteins and other molecules needed for life processes. The simple sugars produced in photosynthesis are also broken down through a process of cellular respiration to provide the energy the plant needs to grow, develop and mature. The products of cellular respiration are energy used by the plant, water and carbon dioxide gas.

The balance between the carbon captured through photosynthesis and the carbon released through respiration determines a plant's growth rate and productivity (Figure 3.1). If environmental conditions support more photosynthesis than respiration, the plant has the energy it needs for normal maintenance, growth and development. If environmental conditions require more energy than it can produce through photosynthesis, the plant will cease to grow and develop normally and will eventually die if it exhausts all available reserves of energy.

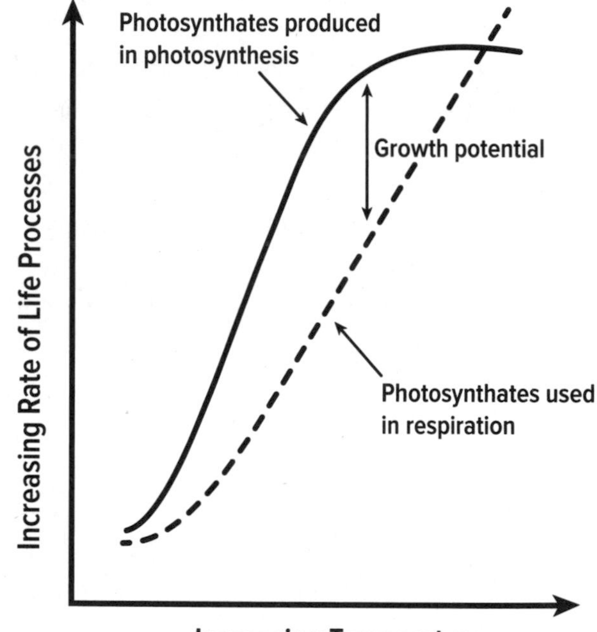

FIGURE 3.1. Plant Growth Potential Linked to Temperature. When the rate of photosynthesis is greater than the respiration rate, plants grow and develop normally. If the rate of respiration is greater than photosynthesis, plants use stored photosynthetic sugars to maintain life processes and become more susceptible to abiotic and biotic stresses. Optimal range in growth temperatures vary among plant species. Temperatures between 54 and 77°F are favorable for cool-season crops like broccoli and wheat, but the growth of warm-season crops like tomatoes, squash and corn is favored by temperatures between about 68 and 77°F. Credit: © Nebraska Extension, University of Nebraska-Lincoln.

Many environmental variables influence the balance between photosynthesis and respiration, but water and temperature are key. Like CO_2, water is a material input to photosynthesis. It is also the primary transport medium for nutrients, proteins, growth hormones and other materials into, through and out of the plant. Temperature governs the rate of photosynthesis and respiration, but these two processes respond independently and often differently to changing temperatures. Optimum temperatures tend to push the balance toward photosynthesis, while temperatures outside of the optimum range tend to favor respiration.

Plants have evolved three different types of photosynthesis processes that have advantages and disadvantages to plant growth depending on environmental conditions. All three processes use water and carbon dioxide plus sunlight to capture and store energy as simple sugars, but they vary in several important ways, including the initial type of sugar produced.[3]

Some plants, called C_3 plants, produce a three-carbon sugar from carbon dioxide. This type of photosynthesis, the most widespread, is used by about 85 percent of all plant species. C_4 plants use a different photosynthetic process to produce a four-carbon sugar. The differences in these two photosynthetic processes have an important consequence: C_4 plants are around 50 percent more efficient at capturing carbon than C_3 plants.[4] C_4 plants are so efficient that increased levels of atmospheric CO_2—essentially an increase in the material input to photosynthesis—does not increase crop production to any great extent. In contrast, C_3 plant photosynthesis becomes much more productive as levels of atmospheric CO_2 increase.

There are some other important differences as well. C_3 plants are most productive at temperatures ranging between about 60°F to 86°F and when ample water is available. C_3 crops are well adapted to temperate climates, but under hot and dry conditions, their photosynthetic sugar production drops dramatically. Most fruit and vegetable crops as well as the cool-season cereal grains—wheat, barley, rye and oats—use the C_3 photosynthetic pathway. Temperature optimums for C_4 photosynthesis are much higher, ranging from 86°F to 113°F, and C_4 plants use water more efficiently, so they can continue to produce sugars in much warmer and drier conditions than C_3 plants. C_4 plants are well adapted to tropical climates. Corn, sorghum, sugarcane and amaranth all utilize the C_4 photosynthetic pathway.

CAM plants are so called because they use crassulacean acid metabolism, a third specialized type of C_4 photosynthesis that is particularly well adapted to extreme desert environments. These plants take up carbon dioxide at night, when it is cooler and more humid, and then carry out C_4 photosynthesis during the day. Pineapple, agave, prickly pear and sisal are all CAM plants in widespread commercial production. There is interest in developing C_3 plants with the ability to utilize CAM photosynthesis as a way to adapt food and biofuel crops to the warmer and dryer conditions that are projected under climate change.[5]

Early research into the influence of climate change on agriculture concluded that crop response to increasing CO_2 levels in the atmosphere would lead to greatly increased yields, because of the "carbon fertilization effect." This perspective viewed the CO_2 levels at the time as a limiting factor to plant growth, particularly in the major food crops like rice, wheat and potatoes, which use the C_3 photosynthetic pathway. Given optimum conditions for plant growth—nutrients, water, temperatures—early research found crop yields increased quite substantially with increasing CO_2 levels.

The carbon fertilization effect is simple, in principle: the more carbon dioxide in the atmosphere the better plants will grow and the more they will yield. Unfortunately, in real-world agricultural production, carbon fertilization is not so simple. Increasing CO_2 tends to have a greater effect on plant growth in dry conditions, and C_3 plants receive more of a yield boost than C_4 plants. Climate change is changing a lot more than just CO_2 concentration. Crop growth and harvestable yield results from the interplay of plant physiology and a myriad interacting environmental factors. Available nutrients and water, nutrient balance, temperature, light, genetics and competition for resources with other crop plants, weeds and pests all come into play.

This early research on carbon fertilization formed the basis for much of the optimism about climate change effects on agriculture; however, as the research matured, it became clear that the benefits of carbon fertilization are often outweighed by the crop stress associated with higher and more variable temperatures, increased variability in precipitation and extreme weather events and other indirect climate change effects such as increased pest pressures.[6]

Temperature

Because plants are mostly rooted in place, they have evolved an exquisite ability to sense and respond to environmental conditions—temperature, light, water and nutrients—to achieve successful growth and reproduction. Plants in the tropics adapted to the challenges of wet and dry seasons, those at the poles to short summers and long, cold winters. Plants in temperate regions adapted to seasonal extremes of winter and summer in two main ways. Perennial plants developed the ability to live many years and go dormant—to pause growth—to avoid the winter or summer extremes, while annual plants evolved to move swiftly from germination to maturity during periods of the year with more moderate conditions. Biennial plants are a special case—annual plants that require a winter pause in order to complete their life cycle.

Temperature plays a special role in plant development, as it does in all biological processes. The rate of plant development—how fast a plant moves through its life cycle from seed germination to seed production at maturity—is determined largely by seasonal temperatures, assuming other factors are not limiting growth. Many critical plant life processes depend on relatively narrow ranges in temperature at specific points in the plant's life cycle. If one of these factors moves out of the optimum range, the plant's growth and development processes are slowed while it diverts resources to resist damage from the associated stress.

Every plant species has a characteristic temperature regime, a set of optimum temperatures that best supports its growth and development from germination to maturity. These temperature optimums change through the phases of the plant's life cycle. For example, the seeds of lettuce, a cool-season crop, will not germinate at soil temperatures below 38°F; however, the optimum range for germination is 43°F to 70°F. Later in the lettuce's life cycle, vegetative growth occurs at air temperatures between 45°F and 77°F, with an optimum between 64°F and 77°F.[7] Table 3.1 shows the temperature regimes of some important annual food crops.

To a certain extent, plants can acclimate to extremes of heat and cold within the limits of the lethal range, as long as the change in temperature is gradual. They have evolved a number of adaptations to manage temperatures outside of their optimum temperature range. With a gradual increase in air temperature, the transpiration rate rises and the increased evaporation at the leaf surface cools the plant. If leaf tissue temperatures continue to increase, leaf stomata close and the leaves wilt, which stops growth but conserves water and protects leaves from the damaging effects of direct sunlight. But high rates of transpiration can exhaust soil water supplies and increase risk of plant damage from drought.

If temperatures gradually cool below the optimum, a common plant response is to slow or stop growth and to accumulate salt-like substances in their tissues. These substances protect the plant from damage as temperatures drop below freezing by lower the freezing point of water held in plant tissues. This process is called "hardening" and is quickly reversed if temperatures rise and growth resumes.[8] Rapid and large changes in temperature that include minimums below freezing, like the extreme variability that is common in late winter and early spring, are more damaging than more gradual changes, because plant acclimation processes cannot respond quickly enough to prevent tissue damage.

For most crops, growth is more rapid as temperatures increase, at least until the upper limit of the optimum range. Although growth tends to decline rapidly once temperatures increase above the optimum, most crops can tolerate high temperatures for short periods through mechanisms that protect their tissues from heat damage and allow critical physiological processes like photosynthesis to continue. Ultimately, high temperatures and heat waves reduce the quality and yield of annual crops and tree fruits as they ripen in the summer and fall and complicate post-harvest handling procedures such as cooling and packing. High temperatures and direct sunlight can sunburn developing fruits and vegetables, sometimes even scalding or cooking them while still on the plant.

TABLE 3.1. Temperature Regimes of Some Annual Food Crops Grown in the United States.

This table reports temperature ranges and optima in degrees Farenheit for germination, growth and reproduction of selected annual food crops. Notice that plant reproductive optima are cooler than the vegetative growth optima and that reproductive failure begins at around 95°F for most crops.

Crop	Germination Soil Temperature (min.°F)	Vegetative Growth (min. °F)	Vegetative (optimum °F)	Reproductive (optimum °F)	Reproductive Failure (°F)
Corn	50	46	77–91	64–72	95
Rice	60	46	91	73–78	95
Wheat	40	32	68–86	59	93
Soybean	50	45	77–98	72–75	102
Potato	40	45	64–77	NA	NA
Onion	37	45	68–77	NA	NA
Tomato	55	53	68–77	59–68	90 day, 70 night
Watermelon	54	59	65–95	68–77	95

NA: Crop is not grown for reproductive structure. Grain Crop Sources: Germination minimums from standard US crop production guides. All other data from Walthall, et al. "Climate Change Effects on US Crop Prodution," Table 5.1, Ch. 5 (p. 61) in Climate Change and Agriculture in the United States: Effects and Adaptation. 2012. USDA Bulletin 1935. Vegetable Crop Sources: Germination minimums from M. Peet and D. Wolfe, "Crop Ecosystem Response to Climatic Change: Vegetables," in *Climate Change and Global Crop Productivity*, K. Reddy and H. Hodges (eds.). 2000. CABI Publishing. "Air temperature data from C. Ramos et. al. Global Change Challenges for Horticultural Systems," in *Crop Stress Management and Global Climate Change*, J. Arans and G. Slafer (eds.). CABI Publishing, except Reproductive Optimum and Failure for Tomato from D. Relf et al. Tomatoes. 2009. Virginia Cooperative Extension Service #426-418 and for Watermelon from Origene Seeds, Watermelon, n.d.

Plant Responses to Seasonal Temperature Patterns: Dormancy and Vernalization

To avoid damage from the seasonal extremes of temperature and moisture, plants and animals have evolved a number of strategies to synchronize growth and development during periods of more moderate conditions. In temperate regions, many annual plants avoid seasonal extremes by completing growth and reproduction within a short period in the spring or fall. Other annual plants have simply adapted to the more challenging summer conditions. Cool season annuals and temperate perennial plants avoid winter extremes and delay flowering until spring has arrived by delaying growth or going into dormancy.

Dormancy describes the winter pause or resting phase that protects a plant's tissues and conserves its energy during extreme winter conditions. In temperate regions, the declining temperatures and shortening days of fall trigger the dormant period and cold temperatures during winter regulate its length. As winter ends and spring approaches, warming temperatures and increasing day length govern spring growth and flowering. Although day length can be an important factor in the growth and development of some crop plants, day length is not affected by climate change and will not be included in this discussion.

Dormancy

As temperatures gradually fall and nights grow longer, growth-inhibiting chemicals in the plant begin to increase while those that promote growth decline. Once the growth-inhibiting chemicals accumulate past a threshold level, plant growth pauses and will not resume even if environmental conditions become favorable. During dormancy, the balance of growth-promoting to growth-inhibiting chemicals changes once again. Growth-inhibitors decline and growth-promoters rise as the hours of cold temperatures accumulate through the winter period. Plants remain dormant until the hours of cold have accumulated to the point that growth is no longer inhibited, at which point it resumes.

The accumulated number of hours of cold temperatures that plants need to break winter dormancy is referred to as the chilling requirement. Temperatures that are just above freezing, usually between 35°F and 45°F, are most effective in meeting the chilling requirements of temperate crops. Hours below freezing do not contribute to the chilling requirement, and hours at temperatures above 70°F will reduce accumulated chill hours. The plant does not have to experience chilling hours consecutively; chill hours just need to accumulate above a specific threshold to break dormancy.

There is a lot of variation in the amount of chill necessary to break dormancy among crop species and as well as within a species. For example, the chilling requirements for apples range from about 100 to

TABLE 3.2. Chilling Requirements for Selected Perennial Fruit Crops. Notice the large variation in chill hours between fruit types and within each type.

Fruit Type	Chilling Requirement (hours between 35 and 45°F)
Apple	100–1100
Pear, European	800–1100
Pear, Asian	600–900
Peach and nectarine	400–1050
Cherry	1000+
Muscadine grape	200–600
Blackberry	50–800
Blueberry, highbush	900–1000
Blueberry, rabbiteye	400–700

Adapted from, A. Powell et al. Fruit Culture in Alabama: Winter Chilling Requirements. 2002. Alabama Cooperative Extension Bulletin ANR-53-D.

1,100 hours, while blackberries range from a low of 50 to more than 800 hours. Table 3.2 reports the chilling requirements for a wide variety of perennial fruit crops.

Vernalization

Winter annual crops like winter wheat, barley and rye require exposure to low temperatures not just to break dormancy, but also to shift from vegetative to reproductive growth and flower in the spring. This type of plant response to low temperature is called *vernalization*. In winter grain crops, vernalization days are used as a measure of accumulated cold temperatures. Like chill hours, vernalization day requirements vary by species and variety. For example, winter wheat typically requires an accumulation of four to six weeks of cool soil and air temperatures ranging between 40°F and 50°F. Once its vernalization requirements are met, the plant then shifts from vegetative to reproductive growth and is capable of flowering. Vernalization requirement is an important criteria for the selection of winter annual crops, as early flowering in grain crops increases the risk of freeze damage, while late flowering increases the risk of damage from heat and drought.

After the chilling requirement is met and dormancy is broken, plant growth resumes as daytime temperatures rise. Temperature now plays a different but equally important role in crop production, because it is the accumulation of warm temperatures and lengthening days following the end of the dormant period that triggers spring flowering.[9]

3.1. Heat Requirement: Growing Degree Hours or Days?

The growth and development of most crops are driven primarily by the accumulation of hours when temperatures fall between about 41°F and 77°F and water is not limiting. Temperatures below this range slow growth processes because photosynthesis is slower in cold temperatures and the photosynthetic sugars that are produced are diverted from growth to defend against damage from cold temperatures. Temperatures above this range slow growth because of an increase in plant respiration and the diversion of plant resources to heat defenses. The threshold of accumulated warmth is expressed in terms of hours for fruit and nut crops and in growing degree days in grain crops.

The rate of plant growth and development is very closely related to the accumulation of warmth over the growing season. This relationship is often used to predict the timing of significant crop development events—like fruit or nut tree flowering or grain fill—using weather data.

In fruit crops, the combination of chilling and heat requirements places a species or variety into the early-, mid- or late-season blooming groups. A crop type with a low chill and a low heat requirement will flower earlier than one with a low chill and a high heat requirement. The combination of chill and heat requirements also determines the risk of crop damage from variable spring weather. For example, figs and some Florida peach varieties both have low chilling requirements (200 to 400 hours), so both crops could be at risk for possible spring freeze damage from early bloom. However, only the Florida peach is at risk from variable spring weather, because it has a low heat requirement and flowers in early spring, while the fig has a very high heat requirement so it will not flower until late spring, greatly reducing the risk of freeze damage. A careful consideration of both the chilling requirement and the growing degree hours of perennial fruit species or cultivars can greatly reduce the risk of crop failure from variable spring weather.

In annual grain crops, plant growth and development are based on an accumulation of growing degree days. In winter annuals, these days control fall and spring plant growth, before and after dormancy. In the summer annual grains that do not go dormant, like corn and sorghum, growing degree days are the main driver of plant development. Cool spring weather lengthens the vegetative growth period and may push the reproductive phase into summer, when yields may be reduced by heat or drought during flowering and grain development. Warmer spring temperatures may cause the plant to speed through the vegetative phase and shift into reproduction before it has had time to grow sufficiently to support high grain yields.

The increased variability in temperatures and a lengthening of the growing season have complicated the management of all crops but present additional challenges to producers growing crops that require a dormant period to reproduce. Warming winters make the prospect of fruit and nut crops failing because they do not accumulate enough chill hours to break dormancy an increasing reality.

3.2. The Decline of Chill Hours in California's Central Valley

Many perennial crops such as fruit trees, nuts and grapes require exposure to a minimum number of chill hours in order to flower and fruit. Researchers interested in understanding how the warming winters associated with climate change might affect fruit production in California's

Central Valley analyzed temperature data for the fifty-year period from 1950 to 2000 and projected temperature trends for later in the twenty-first century using different climate scenarios.[10] They found evidence that California's winter chill declined by as much as 30 percent in some areas between 1950 and 2000. Conservative projections suggest that this decline in chill hours will continue through this century, to 60 percent below the 1950 baseline by mid-century, and by as much as 80 percent below the 1950 baseline near the end of the century. The area capable of producing the highest-quality wines in the Central Valley region is projected to decline by more than 50 percent before the end of the century. Most orchards remain in production for decades and are generally planted assuming that the current temperature conditions will remain constant. The results of this research suggest that growers currently managing young orchards can expect decreasing yields as time goes by, particularly if climate change picks up pace through this century as projected. Planning for new orchards will need to take these projected declines in chill hours into consideration when selecting fruit and nut crop varieties.

Local patterns of temperature—seasonal highs and lows, the length of the frost-free season and differentials between daytime and nighttime temperatures—are important considerations in the production of all crops, but particularly those that enter a period of dormancy. Producers select varieties likely to be productive in their location by matching plant requirements with local climate conditions. Crop planting calendars and crop variety recommendations are based on a body of research conducted to determine suitable crop varieties for specific regions. The bulk of this work was conducted under stable climate conditions and will need to be revised as the pace and intensity of climate changes increase through mid-century. Already there is evidence that increases in average temperatures and variability associated with climate change have reduced crop yields.[11]

Water Availability to Plants

Water availability is often the most limiting factor in agricultural production. Agriculture is only possible where seasonal rainfall patterns are suitable for crop production or there is some means of irrigation or another reliable water supply of adequate volume and quality. All of a plant's life processes are dependent on a continuous flow of water from the soil to the atmosphere along a pathway that extends from the soil into the roots, up the stem to the leaves and out of the leaves through the stomata. This flow, called transpiration, accounts for 90 percent or more of the water taken up by plant roots. Although very little of the water flowing through a plant is chemically bound in plant tissues or actively involved in processes such as photosynthesis, transpiration serves a number of essential functions.[12]

Transpiration moderates plant temperature through the evaporative cooling that takes place on the leaf surface as liquid water is released to the atmosphere. The carbon dioxide needed for photosynthesis diffuses into plant leaves from the atmosphere through the stomata, which are kept open by the transpiration stream. Plants use the transpiration stream as a water-based transport system to move nutrients and other substances into and through their bodies as needed for growth, development and maintenance. A constant stream of water keeps plant cells filled with water so there is adequate pressure for the cells to expand as they grow; this process creates a firmness or fullness in the stems and leaves so the plant remains upright. Transpiration also provides a continuous supply of water for use in photosynthesis.

Crop health, yield and quality reach an optimum when the soil supplies enough water to maintain the transpiration stream throughout the growing season. Plants experience drought stress anytime soil water levels drop too low. Limited soil moisture interrupts the transpiration stream and creates a cascade of effects that reduce crop growth by interfering with many life processes that require a continuous supply of water. Plants have evolved many defenses against variable water supply, because it is common for soil moisture to vary quite considerably with seasonal precipitation patterns.[13] Some shift from vegetative to reproductive growth and completing their life cycle early—though at much reduced seed yield and quality. Others respond to drought stress by making adjustments that reduce water loss and/or increase water uptake. For example, many plants can decrease water loss under drought stress by reducing their leaf area. Common plant responses to drought include wilting, rolling, shedding and/or reducing the number and size of new leaves. Increased root growth is another common drought stress response in many plants. Plants also have various physiological responses to drought stress, producing many different kinds of chemicals that act to conserve cell moisture, encourage the movement of water into cells and prevent or repair tissue damage caused by a lack of water.

An excess of soil moisture can also reduce crop productivity because of the lack of oxygen in the root zone. Many crop plants are very sensitive to waterlogged conditions and can literally drown if these persist for more than 24 hours; however, there is considerable variation in crop tolerance to water-saturated soils. For example, soybeans can survive waterlogged soil conditions for only 48 to 72 hours while corn can survive for two weeks and winter wheat nearly a month.[14]

Most crop plants are well adapted to short periods of waterlogging and water stress, since soil moisture conditions fluctuate quite naturally with precipitation patterns; however, most also have critical growth phases during which an excess or lack of water can result in irreversible damage to developing tissue and greatly reduce crop quality and yield. For example, waterlogging during the germination and seedling growth phases often results in total crop failure, a water shortage early in the crop's development can delay maturity and often reduces yields, and moisture shortages at later growth stages reduce both crop quality and yield. A continuous supply of water is particularly important for the profitable production of fruit and vegetable crops: optimum water availability increases fruit size and weight and prevents quality defects such as tough, cracked or misshapen fruit. Table 3.3 presents the water requirements and critical growth phases of some popular vegetables.[15]

TABLE 3.3: Vegetable Water Needs, Critical Moisture Periods and Defects Caused by Water Deficit.

Notice the wide range in water needs and variation in critical moisture period among vegetable crops. Water needs values are for typical North Carolina summer field growing conditions.

Crop	Drought Tolerance	Water Needs (number of day/inch)	Critical Moisture Period	Defects Caused by Water Deficit	Comments on Moisture Deficit
Snap Beans	LM	7	flowering	poor pod fill and pithy pod	
Carrot	MH	21	seed germination and root expansion	cracked and misshapen roots	
Cucumber	L	7	flowering and fruiting	pointed and cracked fruit	can drastically reduce yield and quality
Leafy Greens (kale, chard, mustard)	L	7	continuous	tough leaves	continuous moisture essential for good yields
Irish Potato	M	7	after flowering	second growth and misshapen tubers	irrigate only during extreme drought during tuber development
Summer Squash	L	5	fruiting	pointed and misshapen fruit	irrigation can double or triple yields
Sweet Corn	MH	14	silking	poor ear fill	irrigation prior to silking has little value
Tomato	M	7	fruit expansion	blossom end rot and fruit cracks	continuous water increases fruit size
Watermelon	MH	21	fruit expansion	blossom end rot	can withstand extreme drought but yields will be reduced

Drought Tolerance: L = low, needs frequent irrigation; M = moderate, needs irrigation in most years; H = high, seldom needs irrigation. Adapted from Vegetable Crop Irrigation, D. Sanders. North Carolina Cooperative Extension Service. Bulletin HIL-33E. 1997.

Indirect Effects on Plants: Weeds, Insects, and Diseases

Climate change has a multitude of indirect effects on crop and livestock production. These are the effects that changing patterns of temperature and precipitation have on the non-crop species found on farms and ranches, such as insects, weeds, disease-causing organisms and invasive species. Climate change effects influence the behavior and productivity of these non-crop species, which, in turn, changes the nature of their relationships with each other and with crops and livestock. Because the relationships between species in agricultural systems are dynamic and the behavior of the system is complex, it is difficult to project how the indirect effects of climate change may influence the performance of the agricultural production system as a whole. The limited information available about how these non-crop species respond to climatic effects may provide some insight on potential impacts.

Weeds

Changes in temperature and precipitation patterns, coupled with increased levels of atmospheric CO_2, change the incidence, population levels and competitive ability of weeds, pests and diseases in the production system.[16] These changes can affect crop yield and quality, increase production risks and may require changes in management or farm system organization. Globally, weeds cause the highest crop losses (34 percent). Insect pests and pathogens are less damaging to yields, causing global yield losses of about 18 and 16 percent respectively.

Temperature and water availability both play important roles in the growth, reproduction and distribution of weeds. Growth and reproductive ability increase as temperature increases within the limits of the characteristic temperature regime of a species. Current weed distributions will likely shift north with warming temperatures, changing the mix of weed species on the farm or ranch as some common weeds leave the system and new weeds enter. More variable water availability will have highly variable effects on weed success, because drought tolerance varies widely among both weed and crop species. The differential response of C_3 and C_4 plants to increasing levels of atmospheric CO_2

adds an additional complexity. A C_3 weed in a C_4 crop will likely become more competitive, while the reverse is likely in a C_4 weed/C_3 crop combination. When the crop and weed share the same photosynthetic pathway, the weed typically benefits more than the crop from increasing CO_2 levels.

Climate change will likely complicate weed management involving cultural, biological and chemical control strategies. Managing weeds with cultivation will be complicated by a combination of warming temperatures and increased rainfall variability. Warming will speed the growth of weeds and narrow potential windows for control because vulnerable growth phases will be shorter, while more variable rainfall will narrow and make more variable the periods when soil conditions permit effective cultivation. Biological controls may be less effective because the synchrony between pest and control agent will likely be disturbed by climatic changes. Chemical controls will also become less reliable because increasing atmospheric CO_2 levels have varying effects on herbicide efficacy. For example, at elevated CO_2 levels, the efficacy of the most commonly used herbicide in the United States (glyphosate) was unchanged for pigweed but was reduced for lambsquarters, Canada thistle and quackgrass. As a general rule, farmers and ranchers can look to comparable production systems to the south or in regions where current climate conditions are similar to projected conditions in their region for insight about troublesome weed challenges ahead.

Insects

Many of the observations about climatic effects on weed success hold true for insect pests.[17] Warming temperatures tend to increase pest success because accelerated life cycles reduce time spent in vulnerable phases of the life cycle. For many pests, it also appears that temperature increases to date have reduced stress from cold temperatures without increasing stress from maximum temperatures, increasing winter survival and promoting a northward expansion of range. These changes tend to increase pest populations, change the mix and relative numbers of existing populations and facilitate the introduction of novel insects to the farm.

Longer growing seasons change several factors that tend to increase the potential for insect pest damage to crops and livestock. Pest populations increase as seasons lengthen because pest species can produce more generations in a growing season. Higher pest populations over a longer time increase the potential for pest damage to crops and livestock as well as the development of pest resistance to management strategies.

Management of insect pests will almost certainly become more challenging and costly as pest populations increase, generation times decrease and ranges expand northward. The change in pesticide use for the successful chemical control of earworm in sweet corn along the East Coast provides some insight into coming challenges. Sweet corn is treated for corn earworm from 15 to 32 times per season in Florida, 4 to 8 times in Delaware and 0 to 5 times in New York. Changing climate is likely to challenge existing cultural and biological pest management strategies as well. Like weed management, a look at comparable production systems to the south provides a glimpse of future pest management challenges.

Diseases

Climate change effects on crop and livestock losses from disease are similar to those for weeds and insects, but pathogen-host interactions add an additional level of complexity.[18] Changes in seasonal patterns of temperature and precipitation, warming temperatures, more extreme weather events and increasing atmospheric CO_2 levels alter the geographic ranges and relative abundance of pathogens, increase their rates of spread and evolution and influence the effectiveness of host resistance to infection. More variable weather may increase or decrease the potential for infection depending on the temperature and moisture requirements of host and pathogen and how these overlap. In addition, temperature and moisture stress can lower crop and livestock resistance to disease. When a combination of effects that result from increased temperature, for example, are no longer ideal for the host, losses to productivity can be compounded when the change in temperature coincidently favors the disease organism's success.

Like weed and insect management, disease management is likely to become more difficult and costly under climate change. Changes in the timing, spread and infectivity of disease organisms will complicate cultural and biological management and resistance to chemical controls will emerge more rapidly. An increase in the assessment and development of novel crops and livestock in an effort to reduce climate risk to existing production systems will increase the risks of new disease introductions and may create new opportunities for existing diseases as well.

Livestock Sensitivities

Environmental conditions play a crucial role in livestock behavior, development and productivity.[19] Like plants, animals have temperature and water requirements that support healthy growth and development, and sensitivities to climate effects vary over their life cycle. Temperature and water requirements vary by species and breed, are influenced by physical traits (for example, coat color or thickness) and change throughout the animal's life cycle (for example, young animals are most vulnerable to extremes in temperature). Unlike plants, animals have some ability to regulate environmental stresses by changing location in response to challenging conditions. For example, animals often seek shade as temperatures rise or move to sheltered areas during high winds or heavy rains.

Because well-managed livestock are typically provided adequate water and food, temperature and humidity are the key sensitivities in livestock production systems.[20] The optimal temperature ranges for livestock species allow the animal to maintain a relatively constant core body temperature without significantly altering behavior or physiological function. Outside of this range, disruptions in fertility and growth can reduce productivity and even cause death. Like plants, animals can acclimate to gradually changing temperatures, but long periods of temperature extremes, for example heat waves and cold waves, and extreme fluctuations in temperature can reduce productivity and also result in death of the animal. Combinations of wet conditions, cold temperatures and wind during the spring birthing period, a particularly sensitive time in the life of an animal, can present significant challenges to producers managing livestock outdoors.

Temperature is a particularly critical environmental variable. Animals can tolerate a relatively wide range of fluctuations in temperature; however, if environmental conditions become too extreme, they undergo a physiological shift to maintain core body temperature at the expense of normal growth and development. This shift is often reflected in behavioral changes—such as a decline in physical activity, reduced feed and water intake and slowed digestion—and explains why livestock productivity tends to decline during periods of temperature extremes.

The potential for loss of livestock productivity as a result of environmental extremes depends on the interaction of a number of factors. Management factors like prior condition, stocking rate, nutrition and type of production facilities interact with environmental factors such as the rate and extent of change and the persistence of the extreme conditions to determine productivity effects of environmental changes.[21] For example, US beef cattle are typically held in extensive outdoor facilities, because they can cope with most environmental conditions as long as these only change gradually and extreme events are of short duration. In contrast, swine and poultry are typically produced in controlled environment facilities in which air flow and temperature are controlled to buffer outside environmental extremes; however, if the facility is not designed to cope with extreme events or if energy or water supplies are interrupted during an extreme weather event like a heat wave, conditions can quickly become lethal.

Grazing Lands and Forage Crops

Grazing lands, rangelands, pastures and forage crops are managed to produce food for livestock. Rangelands are extensively managed natural grasslands, most of which are found in the western United States. These are often federal lands leased by livestock producers, though many privately owned ranches in the West manage large tracts of native rangelands. Pastures and forage crops are intensively managed and are typically located in areas of abundant rainfall or where sufficient water is supplied through irrigation. These lands are usually privately owned and are planted with improved varieties of grasses and legumes that are adapted to local soil and climate conditions. Pastures are typically grazed by livestock, while forage crops are harvested, processed and fed to livestock. Achieving high productivity of livestock produced on grazing lands or fed forages requires the manager to carefully match grassland and forage crop quality with the nutritional needs of the animals in each phase of the livestock life cycle. For example, nursing cows must have access to adequate amounts of high-quality grasslands or forages in order to create the nutritious milk needed to produce healthy and well-grown calves.

The productivity of grasslands and forages are regulated by the interaction of seasonal patterns of temperature and moisture with local soil conditions.[22] Given sufficient soil moisture, either from rain or irrigation water, higher temperatures tend to increase grassland and forage production as long as temperatures remain within a productive range; however, if water is limiting, high temperatures can reduce yields. Increased concentrations of CO_2 in the atmosphere complicate temperature and soil moisture interactions on grassland and effect forage productivity. Grassland and forage plants respond directly to an increase in CO_2 through two different mechanisms that both increase yield. At higher concentrations of CO_2, plant stoma partially close, which reduces water loss from the plant and helps to maintain growth at higher temperatures and lower soil moisture levels. Higher CO_2 concentrations also increase yields by directly stimulating increased photosynthesis, particularly in C_3 grassland and forage species. Essentially, higher CO_2 concentrations have a sort of "irrigation effect," by improving forage yield under moderate drought conditions.

Both increases and decreases in the nutritional quality of grasslands and forages have been observed in response to rising temperatures and CO_2 levels.[23] If soil nitrogen levels remain the same or decrease as temperatures rise and CO_2 levels increase, protein content can be reduced if the changed conditions stimulate plant growth. Rising temperatures and CO_2 levels can also change the mix of species present in the grassland, increasing or decreasing forage quality.

Climate change can also indirectly effect livestock production through changes in the quality and supply of feed grains and the increased incidence of livestock pests and diseases. Changing climate conditions and increasing CO_2 concentrations can reduce grain production and total grain protein content. Warmer and more humid conditions may favor some pathogens and parasites that reduce livestock productivity and will facilitate the expansion of these diseases.

Natural Resource Sensitivities

Agriculture is a complex biologically regulated system that is linked closely to climate through the direct and indirect effects of changing temperature and precipitation patterns coupled with increasing concentrations of atmospheric CO_2. The ecosystem within which a farm or ranch resides supports the production of crops and livestock through biological processes that capture, store and recycle solar energy; regulate soil quality, crop nutrient release and the quality and quantity of water in the landscape; and provide pest suppression, pollination, biodiversity conservation and carbon sequestration services.[24] All of these crucial ecosystem services are sensitive to changing climate conditions.

Soil Resources

Along with climate, soils determine the natural productive capacity of a farm or ranch system. Soils are the living foundation for the agricultural ecosystem, or *agroecosystem*, and provide a number of services that support crop and livestock production.[25] Soils supply all the water and most of the nutrients that plants need for healthy growth and development. Healthy soils suppress weeds, insect pests and diseases and produce chemicals that promote plant growth, degrade toxins and recycle plant and animal wastes. The living soil, just like plant and animal life, is responsive to temperature and moisture conditions and can be stressed when conditions exceed optimum ranges. When climate change effects interfere with healthy soil life processes, the ecosystem services provided by soils can be significantly reduced.

Soils can be damaged by extremes in temperature and precipitation, particularly if the surface is exposed to the elements. Soil can be picked up and carried away by wind or water during violent storms. Carrying out field operations like planting, cultivation and harvest when the soil is too wet or too dry can damage and degrade soil. Longer growing seasons and warmer winters increase the nutrients plants release during periods of low or no crop nutrient demand, which can increase the potential for nutrient loss into surface and groundwater, particularly during periods of heavy rainfall.

Changing climate conditions have already begun to complicate soil management and degrade soil and water quality on farms. The removal of topsoil and nutrients and the addition of pesticides from agricultural soils via wind and water also have far-reaching effects on environmental quality, both locally and globally.

Water Resources

Water is a major limiting factor to agricultural productivity, and water resource management in a changing climate is widely recognized as the most critical near-term resource sensitivity for agriculture.[26] More frequent and intense extremes of temperature and precipitation are already creating significant disruptions in water supplies throughout the country. These climate effects are predicted to increase in frequency and intensity in coming years.

Rising temperatures, reductions in snowpack and shifting precipitation patterns have already begun to alter the patterns of agricultural demand for water as well as the availability and cost of water throughout the United States.[27] Competition for water among agricultural, industrial and public uses has already begun to reduce water supplies to agriculture in the Southwest, and a similar rivalry has even begun in the relatively water-abundant Southeast. These conflicts are projected to intensify through mid-century and will likely result in significant reductions in water supplies for agriculture in the major regions of irrigated agriculture where much of the nation's fruits, vegetables, beef and poultry are produced.

Water quality is also likely to be degraded by the changing climate through a number of pathways. The increase in number and intensity of heavy rainfall events is likely to increase soil erosion and runoff into surface waters, while warmer winters are likely to increase the loss of nutrients to groundwater. Both changes will increase the nutrient, chemical and sediment loads in surface waters and nutrient loading in groundwaters. Increased pest pressures will likely require increased use of herbicides and pesticides, which will contaminate sediments leaving the farm or ranch in runoff waters. And warmer temperatures, coupled with more frequent and intensive drought and heat waves, will increase surface water temperatures and evaporation, concentrating pollutants and further degrading surface waters.

Changing climate conditions have already begun to complicate on-farm management of water and reduce the quantity and quality of water resources used for agriculture in many areas of the country. These impacts are projected to increase in intensity through mid-century.

Human, Built and Financial Resources

Farm or ranch resources that are most closely connected with the human aspects of agriculture are important considerations when assessing the sensitivity of the production system. The response of management and labor to changing climate conditions, the potential damages to infrastructure from climate change effects, and the financial well-being of the production system all influence sensitivity to climate change effects.

Climate change will likely increase the challenges of farm and ranch management, because of the greater uncertainties and increased complexity it will bring. New pests, novel crop and livestock behavior, shorter windows for timely fieldwork, more variable yields, increased competition for water, new regulatory requirements and changing markets all complicate production. In addition, the increased need to innovate with new crops, livestock, management strategies, tools and technologies adds a novel complexity to farm management. Further, the effects of extreme weather, particularly heat waves and droughts, on fieldworkers will make it difficult to complete field operations in a timely manner and retain good employees. Shifts in crop and livestock growth patterns and sudden changes in production conditions can disrupt the flow of farm and ranch work, making it difficult to maintain a consistent labor force. If current production system conditions contribute to a highly stressful environment, the additional challenges presented by changing climate conditions may represent a key sensitivity of the farm production system.

Climate change effects, particularly extreme events like violent storms or heavy precipitation that causes flooding, present specific hazards to infrastructure such as buildings used for plant and animal production and the processing and storage of products as well as storage for feed, forage and farm equipment and fencing, roads, water and irrigation systems. Lack of access to or damage or loss of these structures and their contents may represent key sensitivities, depending on the importance of the built resource to the production system and the timing of the weather event.

Climate risk is likely to increase operating, maintenance and overhead costs in many production systems. The increased variability of temperatures and precipitation and the increased number and intensity of extreme events will likely result in more instances of reduced product quality and yields, production failures, loss of markets and increased input and insurance costs. New costs will likely be incurred for climate change adaptations and the increased costs related to the maintenance, repair and replacement of crops, livestock and infrastructure subject to climate hazards. The maintenance of increased recovery reserves will also add to the costs of production. These increased costs represent key production system sensitivities, particularly if current farm profitability is low and recovery reserves are inadequate.

Potential Impact Is a Function of Exposure and Sensitivity

The potential impact of climate change is defined by the unique interaction of the production system with the climate system at a specific place and time. An understanding of the key sensitivities as they relate to likely climate exposures provides useful information about the potential impacts of climate change on the farm or ranch that can motivate, inform and guide actions to reduce sensitivities. An additional strategy for reducing the impact of climate change on agricultural production systems is to take action to enhance the adaptive capacity or resilience of the system, the subject of the next chapter.

Endnotes

1. Adapted from Climate Impacts Group, University of Washington, King County, and Washington, ICLEI—Local Governments for Sustainability. Conduct a Climate Change Vulnerability Assessment. Ch. 8 in *Preparing for Climate Change: A Guidebook for Local, Regional and State Governments*. 2007. cses.washington.edu/cig/fpt/guidebook.shtml
2. A threshold is the point at which there is an abrupt change in a quality or property of a system, or where small changes in one or more external conditions produce large and persistent responses in the system.
3. S. Gliessman. The Plant. Ch. 4 in *Agroecology: The Ecology of Sustainable Food Systems*. 2007. CRC Press: Boca Raton, FL.
4. R. Sage and X. Zhu. 2011. Exploiting the Engine of C_4 Photosynthesis. *Journal of Experimental Botany*, Vol. 62, No. 9, pp. 2989–3000.
5. For example, see L. Escamilla-Treviño, "Potential of Plants from the Genus Agave as Bioenergy Crops." 2012. *BioEnergy Research*. Vol. 5 (1): 1–9.
6. C. Wathall et al. "Climate Change Science and Agriculture." Ch. 4 in *Climate Change and Agriculture in the United States: Effects and Adaptation*. 2012. USDA Technical Bulletin 1935.
7. Soil temperature data from M. Peet and D. Wolfe, Crop Ecosystem Response to Climatic Change: Vegetables, in *Climate Change and Global Crop Productivity*, K. Reddy and H. Hodges (eds.). 2000. CABI Publishing. Air temperature data from C. Ramos et al., Global Change Challenges for Horticultural Systems, in *Crop Stress Management and Global Climate Change*, J. Arans and G. Slafer (eds.). 2011. CABI Publishing.
8. FAO. 2005. Frost Protection: Fundamentals, Practices and Economics. UN-Food and Agriculture Organization.
9. Adapted from A. Powell et al., Fruit Culture in Alabama: Winter Chilling Requirements. 2002. Alabama Cooperative Extension Bulletin ANR-53-D.
10. E. Luedeling, M. Zhang, and E. H. Girvetz. 2009. Climatic Changes Lead to Declining Winter Chill for Fruit and Nut Trees in California during 1950–2099. PLoS ONE 4(7): e6166. doi:10.1371/journal.pone.0006166plosone.org /article/info:doi/10.1371/journal.pone.0006166
11. J. Hatfield et al. 2014. Agriculture. Ch. 6 in *Climate Change Impacts in the United States: The Third National Climate Assessment*, J. Melillo, T. Richmond, and G. Yohe (eds.). U.S. Global Change Research Program, 150–174. doi:10 .7930/J02Z13FR.
12. D. Holding et al. *Plant Growth Processes: Transpiration, Photosynthesis, and Respiration*. 2013. U. of Nebraska Cooperative Extension.
13. M. Farooq et al. Plant Drought Stress: Effects, Mechanisms and Management. 2009. *Agron. Sustain. Dev.* 29: 185–212.

14. J. Bailey-Serres et al. Water Proofing Crops: Effective Survival Strategies. 2012. *Plant Physiology* Vol. 160: 1698–1709.
15. D. Sanders. Vegetable Crop Irrigation. 1997. North Carolina Cooperative Extension Service. ces.ncsu.edu/hil/hil-33-e.html
16. C. Walthall et al. Indirect Climate Effects: Weeds and Invasive Plant Species, Climate Change Science and Agriculture. Ch. 4 in *Climate Change and Agriculture in the United States: Effects and Adaptation*. 2012. USDA Technical Bulletin 1935.
17. C. Walthall et al. Indirect Climate Effects: Insects, Climate Change Science and Agriculture. Ch. 4 in *Climate Change and Agriculture in the United States: Effects and Adaptation*. 2012. USDA Technical Bulletin 1935.
18. C. Walthall et al. Indirect Climate Effects: Pathogens, Climate Change Science and Agriculture. Ch. 4 in *Climate Change and Agriculture in the United States: Effects and Adaptation*. 2012. USDA Technical Bulletin 1935.
19. C. Walthall et al. Grazing Lands and Livestock: Climate Change Effects on U.S. Agricultural Production. Ch. 5 in *Climate Change and Agriculture in the United States: Effects and Adaptation*. 2012. USDA Technical Bulletin 1935.
20. Ibid.
21. Ibid.
22. Ibid.
23. Ibid.
24. Walthall et al. Agricultural Soil Resources: Climate Change Effects on U.S. Agricultural Production. Ch. 5 in *Climate Change and Agriculture in the United States: Effects and Adaptation*. 2012. USDA Technical Bulletin 1935.
25. Ibid.
26. Walthall et al. Agricultural Water Resources and Irrigation: Climate Change Effects on U.S. Agricultural Production. Ch. 5 in *Climate Change and Agriculture in the United States: Effects and Adaptation*. 2012. USDA Technical Bulletin 1935.
27. Walthall et al. Water Supply Availability: Climate Change Effects on U.S. Agricultural Production. Ch. 5 in *Climate Change and Agriculture in the United States: Effects and Adaptation*. 2012. USDA Technical Bulletin 1935.

CHAPTER 4

Understanding Adaptive Capacity

A record-setting July heat wave in California in 2006 causes the death of 70,000 poultry and more than 25,000 dairy cows.[1] Farmers across the Southeast suffer widespread losses during severe drought conditions in 2007.[2] High rainfall in June 2008 causes massive flooding throughout the Midwest.[3] Hurricanes Irene and Lee become the largest and most expensive natural disaster in New York State history when they hit within one week of each other in 2011. A late spring freeze following an unusually warm spring makes 2012 the worst year ever recorded for Michigan fruit growers,[4] while 80 percent of agricultural land in the United States suffers under drought conditions resulting in a record high 14 billion dollars in crop insurance payments.[5] US beef prices rise to record highs in 2013 as a result of a historic drought in Texas cattle country the previous year.[6] And in spring 2014, for the first time in the state's history, California vegetable and fruit growers learn that they will get no water from federal or state water projects as water supplies dwindle due to warmer winters and a continuing extreme drought.[7]

Many of the farmers and ranchers profiled in this book experienced one or more of these extreme events, but the impact on each was different. The *potential impact* of climate change on a farm or ranch arises from the interaction between specific *exposures*—the climate change effects experienced at the specific location—and the particular *sensitivities* of the production system to those exposures.[8]

Richard de Wilde credits his local community and his farm's 1,500 CSA members with providing crucial support that allowed him to recover from two back-to-back 2,000-year flood events that ended the 2007 growing season two months early and destroyed early summer crops in 2008 on his farm in southern Wisconsin. In 2011, Jim Hayes' farm in south-central New York was right in the path of hurricanes Irene and Lee. The catastrophic flooding caused by the storms shut off power, destroyed roads and damaged many private homes in the region, causing severe disruptions in Jim's direct market sales that led to a 30 percent drop in income for more than two years. Jim Koan lost 90 percent of his apple crop to the 2012 freeze in Michigan, but he maintained cash flow in 2012 through the sales of excess inventory held over from an abundant 2011 crop that he fermented and bottled. Paul Muller has enjoyed the proven water-conserving benefits of cover crops on his farm in California for more than thirty years, but three years of extreme drought and new FDA food safety regulations that conflict with local water recycling programs have created new water management concerns.

The *vulnerability* of the farm or ranch system to climate change is a function of the *potential impact* and the *adaptive capacity* of the system—the ability of the system to cope with or adjust to climate effects, moderate potential damages and take advantage of opportunities created by climate change. For example, historic drought in Texas in 2012 drove many cattle producers out of business because their land could not produce enough feed or water to maintain existing herds. In other words, the adaptive capacity of industrial ranching systems in the region was exceeded under the extreme weather conditions of 2012. In contrast, the high-quality soils and native grasslands on the 77 Ranch had the capacity to maintain Gary Price's cattle herd during the record drought. The drought conditions did not exceed the adaptive capacity of the improved natural resource quality—soil, water and grassland—produced by the planned grazing and prairie restoration practices in use on the 77 Ranch, and Gary had cattle to sell when prices rose as the cattle supply declined in 2013.

The adaptive capacity of a farm or ranch emerges from the producer's management of five basic agricultural resources or assets—human, social, natural, physical and financial—to achieve specific production goals. Broadly speaking from this asset management point of view, the adaptive capacity of industrial models of production arises from the management of large land holdings (typically owned plus rented land), purchased inputs (e.g., irrigation, fertilizer, pesticides) and government subsidies (e.g., education, research, development and extension, direct payments and insurance, agricultural labor exemptions) to produce commodity products regardless of local resource conditions.

In contrast, the adaptive capacity of sustainable models of production arises from the management of smaller land holdings (typically owned), production inputs produced by healthy soils and agrobiodiversity (e.g., natural precipitation, crop nutrients released by decomposition, pest suppression by beneficial insects) and social capital (e.g., direct markets, community-based research and education) to produce high-value food products that are well adapted to local resource conditions. The choices that agricultural producers routinely make about the assets they manage—people, land, crops, livestock, infrastructure and finances—determine in large part the ability of the farm or ranch to sustain production under challenging climate conditions.

Over the last fifty years or so, industrial agriculturalists have emphasized technological and financial subsidies to enhance the adaptive capacity of their farms and ranches to weather and market shocks, while sustainable agriculturalists have emphasized natural and social subsidies to do the same. While these different strategies offer some climate risk benefits, the adaptive capacity of both industrial and sustainable production systems could be enhanced by equal access to all asset types and to a greater diversity of options within each asset type.[9]

Returning to an earlier example, industrial cattle producers could enhance the adaptive capacity of their production systems to climate change effects by cultivating natural and social assets in addition to technological and financial ones. Rather than limiting financial subsidies to those that lower feed prices and provide disaster relief to industrial cow-calf producers, such subsidies could be redirected to enhance natural resource assets such as forage-based sustainable beef production practices. Likewise, improved access to technological and financial assets could enhance the adaptive capacity of sustainable cattle producers.

Another useful framework for understanding adaptive capacity is to consider how the three different components of adaptive capacity—the operating context, the individual capability to act and existing knowledge and options—interact to create unique place-based opportunities to enhance the adaptive capacity of any particular farm or ranch.[10]

4.1. Enhancing Adaptive Capacity with Diversified Asset Management

Farm and ranch families are typically involved in a number of activities to make a living, both on and off the farm or ranch. This might include individual family members taking responsibility for managing different enterprises, or the whole family focusing their efforts on one enterprise. In addition to farm or ranch work, it is common in US agriculture for at least one family member to contribute significantly to household income through an off-farm job. These family livelihood activities are continuously evolving in response to changing access to the human, social, natural, physical and financial assets described in the Table 4.1.

Careful management of these assets is widely accepted as a normal part of good agricultural business practice, although they are sometimes grouped a bit differently and may be referred to as capital, assets or resources. Farmers and ranchers are experienced risk managers, making tradeoffs between goals and available assets to maintain the performance of their farm or ranch in dynamic production and marketing conditions that are often influenced by weather variability and extreme events; however, few have direct experience managing these assets to cultivate adaptive capacity to climate change effects.

TABLE 4.1 Understanding Adaptive Capacity in Rural Communities

Community Characteristics	Description
Flexibility	The level of social, cultural, political, economic and environmental flexibility within a community
Capacity to Organize	The formal and informal capacity to develop adaptation plans, establish and maintain community organizations, make decisions and manage community-based recovery responses following extreme climate events
Capacity to Learn	The extent to which the capacity to organize is moderated by adaptive learning and subsequent modification of adaptation plans
Assets Available to Management	**Description**
Natural	The productivity of land and actions to sustain or enhance the quality of the natural resources upon which agriculture depends
Human	The capacity of individuals to contribute to the productivity of labor and manage natural resource-based businesses
Social	Reciprocal claims on others by virtue of social relationships, close social bonds that facilitate cooperative action, formal and informal social linkages that facilitate the sharing of ideas and resources
Physical	Infrastructure, equipment and technologies produced with economic activities supported by other types of capital; the built environment
Financial	The level, variability and diversity of income sources and access to other financial resources (credit, savings) that contribute to wealth

Adapted from Table 15.1, N. Marshall et al. 2009. "Enhancing Adaptive Capacity," Ch. 15 in C. Stokes and M. Howden (ed.), *Adapting Agriculture to Climate Change: Preparing Australian Agriculture, Forestry and Fisheries for the Future.* CSIRO Publishing: Collingwood, Victoria, Australia.

Although decisions about how best to manage climate risk ultimately lies with individual managers, community characteristics create barriers and opportunities for effective adaptation. Four essential community characteristics appear to explain why some rural communities are better able than others to adapt to the uncertainties associated with climate change: flexibility, the capacity to reorganize, the capacity to learn (described in Table 4.1) and an asset base that offers a diversity of options within each asset type and equal access to all types of assets.[11] Government policies and programs, agricultural businesses, agricultural education, research and development institutions and community-based organizations all play an important role in creating the community conditions that enable farmers and ranchers to enhance the adaptive capacity of US agriculture.

The Operating Context

Farms and ranches are located in a specific place and are subject to a unique combination of ecological, social and economic conditions. These conditions interact with each other and the production system to create an operating context unique to each farm or ranch. Ecological conditions such as local topography, seasonal weather patterns, climate and the quality of natural resources on the farm or ranch—particularly soil and water—present challenges and opportunities that shape the choices producers make about the kinds of crops and livestock to produce and how to produce them. Social and economic conditions both on and off the farm or ranch such as the quality of family relationships and family business and personal goals, government regulation and support programs, access to financial and technical resources, marketing opportunities and community support for agricultural businesses interact with ecological conditions to influence the design and management of agricultural production systems. Within the limits created by these ecological, social and economic realities, farmers and ranchers can identify challenges and opportunities for enhancing the adaptive capacity of their production system to climate change effects.

Ecological Conditions

The ecological context within which the farm or ranch resides plays a crucial role in shaping the adaptive capacity of the production system. Because agricultural production is based on the availability and quality of natural resource assets, particularly soil, water and climate, natural resource conditions present significant opportunities and challenges to the management of adaptive capacity on farms and ranches.

Competition for water is a good example of the profound effect that disruptions in the natural resource base can have on agricultural systems. In much of the western part of the United States, natural rainfall is too low for the production of many cultivated crops and improved pastures. Over the last century, agricultural production systems dependent on irrigation evolved in this region with the support of favorable agricultural policies that subsidized the development of irrigation infrastructure and the delivery of water supplies. Unsustainable water use by agriculture and industry, unsustainable development policies and rapid population growth in the region's metropolitan areas have degraded soil and water resources and reduced the supplies of water available to agricultural production. Continued access to sufficient quantities of high-quality water is a key ecological condition that shapes the operating context of agriculture in the southwestern United States.[12]

But water is just one of many natural resources upon which agriculture depends. Regional ecosystems provide a suite of natural resource services to the farms and ranches that reside within them. The regulation of soil quality, water quality and quantity, waste processing, the suppression of pests, pollination and climate protection (e.g., flood control, wind and temperature moderation) are just a few of the ecosystem services that directly influence the operating context of a production system.[13] These natural assets produced by ecosystem services on farms and ranches and in their surrounding communities are increasingly compromised by activities that degrade ecosystem health such as pollution from urban, suburban and rural areas; the intensive use of agricultural chemicals and fertilizers, overgrazing grasslands, overdraft of ground and surface waters and the loss of biodiversity.

Climate change represents an emerging threat to ecosystem health that is projected to intensify over this century. Increasing temperatures, longer growing seasons, warmer winters, more variable weather and the increased frequency and intensity of extreme weather events have the potential to amplify existing ecosystem degradation processes associated with agricultural landscapes such as the degradation of soil and water quality caused by soil erosion during heavy rains, disruptions in pollination services and the environmental impacts of increased fertilizer and pesticide use.[14]

Social and Economic Conditions

Social and financial conditions play an important role in shaping the operating context of US farms and ranches.[15] The cost and availability of basic agricultural inputs—land, energy, water, seed, fertilizer and labor—as well as access to profitable markets and changing consumer demands are some social and economic conditions that commonly shape decisions made by producers. For example, federal policies designed to keep agricultural commodity prices low, often below the real costs of production, have played a significant role in the decisions made by many US farmers and ranchers to specialize in one or two commodities and to expand their production volume by increasing acres under management. Government regulations that direct the availability and use of agricultural inputs, as well as product handling and marketing, are a consideration in production and marketing decisions, as are government programs that address the environmental impacts of agricultural operations.

Individual Capability

In many parts of the United States, climate change has increased the complexity of agricultural management because of the novel uncertainty created by increased weather variability and more frequent and intense extreme events. These effects are projected to increase in frequency and intensity as climate change continues through mid-century and beyond. Effective climate risk management will depend on human capacity, both on the farm or ranch and in the local community, to develop new abilities to learn, plan and adapt to changing climate conditions. Taking action to enhance the adaptive capacity of the production system will likely make new and different demands on farm and ranch managers, may increase production costs and will likely require a high level of financial and emotional flexibility.[16]

The motivation of individuals to take actions that enhance the adaptive capacity of their farms or ranches is influenced by their perceptions of climate change.[17] Varying perceptions of climate risk, combined with the geographic variability in climate change effects and differences in production system sensitivities to climate effects, all play a role in individual motivation to take adaptive action. Ultimately, the successful management of agricultural production systems under changing climate conditions will be determined by the willingness and ability to take action to reduce climate risk. Once a producer is motivated to take action, access to existing knowledge and options for effective adaptive action is the final component of adaptive capacity.

4.2. US Producers' Perceptions of Climate Change

Recent research has documented the ability of rural people engaged in agriculture to detect changes in local climate, such as altered plant and animal phenology, new distributions of species, shorter or longer growing seasons and a shifting frequency of extreme weather events. The changes that people notice tend to be closely related to the aspects of the weather that have the most direct effect on their livelihoods, to such an extent that residents of the same community may identify different changes depending on their occupation. Other factors that influence perceptions of local climate change include recent personal experience of extreme weather events, which can heighten the perception of climate risk and pre-existing beliefs and attitudes about global warming that lead people to selectively remember weather events in ways that reinforce their worldview. Biased perceptions of local climate changes influence individual capability to act and present barriers to effective climate change adaptation.[18]

The perceptions and concerns of US farmers about climate change issues have only recently begun to be explored. Here are the results of three surveys: two large surveys conducted in 2011 and 2012 of industrial farmers growing commodity crops in the Midwest and Southern regions of the country, and focus group research conducted with farmers in the state of Maine growing seven major commodity crops.

Sixty-eight percent of 1,242 corn and soybean farmers in the Midwest and Northern Great Plains regions reported that they believed in climate change. Of this group, farmers who believe that climate change is caused by human activity (8 percent) expressed higher levels of concern about likely impacts than those that believe climate change is caused by natural factors (25 percent) or a combination of natural and human factors (33 percent). Surveyed farmers also identified the greatest weather-related challenges to their farms. Longer dry periods, drought and excessive heat headed the list of concerns, followed by increased pests and diseases and excessive rainfall. Almost 60 percent of the farmers surveyed believe that they should take steps to protect their land from increased weather variability.

A survey of more than 1,000 commodity crop producers growing corn, cotton, grain sorghum, rice and wheat in Mississippi, Texas, North Carolina and Wisconsin discovered that 40 to 50 percent do not believe that climate change has been scientifically proven, while the remaining farmers split about evenly between those that do believe it has been proven and those having no opinion. Between 60 and 75 percent of respondents believe that normal weather cycles explain most or all of the recent changes in climate. Farmers reported that they would be most likely to diversify crops, buy crop insurance, modify lease arrangements or exit farming altogether if weather variability and extremes intensify.[19]

About 150 Maine farmers representing seven major commodity groups—potato, dairy, blueberry, vegetable, apple, beef and nursery plants—participated in focus groups to discuss the future of agriculture in their state. When asked about future challenges, only one farmer mentioned climate change and only a few farmers, primarily blueberry and apple growers, expressed concerns about changing environmental conditions; however, most of the participating farmers expressed concerns about new challenges associated with current weather changes such as more erratic weather, new pests and more extreme weather events. Farmers reported a variety of adaptive actions taken in response to these changes such as crop diversification, adding drainage or irrigation systems, season extension in protected growing conditions such as hoophouses, and ecological production practices that build soil quality to improve cropping system capacity to buffer weather extremes.[20]

Existing Knowledge and Options

Production risks associated with weather variability and extreme events have always been an important consideration in agricultural management. Climate change has not changed the nature of those risks; however, projected increases in the frequency and intensity of weather variability and extremes are expected to create unprecedented production challenges. The adaptive capacity of a production system depends on the knowledge and technologies available to support human capacity to manage climate risk—the knowledge and physical assets available to the production system. Sustainable agriculture, adaptive management and resilience thinking offer a suite of principles, practices, tools and technologies uniquely suited to designing and managing agricultural production systems with a high adaptive capacity to climate change effects.[21]

Sustainable agriculture and agroecology, the science of sustainable food systems, provide a strong foundation upon which to build a resilient US agriculture.[22] A rich knowledge base of practical application and innovation, informed by a deep understanding of the ecology of agriculture, can be used to guide the development of locally adapted, sustainable and climate-resilient systems of agricultural production. Over the last forty years, farmers, ranchers and others in the United States committed to sustainable food systems have worked together to develop and share information about agricultural sustainability through both formal and informal research, teaching and learning networks.[23] This accumulated knowledge offers the opportunity to apply a wealth of conceptual understanding, well-tempered by practical, place-based experience, to enhance the adaptive capacity of US agriculture and food systems increasingly challenged by resource scarcity and climate change.

Key sustainable agriculture principles that have proven useful in the management of production risks related to weather variability and extremes include an emphasis on soil quality, planned biodiversity, ecological design and sales into high-value markets. Healthy soils are a key climate risk management tool, acting to buffer the farm or ranch from increased variability and the extremes in precipitation being experienced across the United States. Planned biodiversity, such as diverse crop rotations and the integration of crops and livestock, spreads climate risk through the growing season, reducing potential losses from any single weather event. Ecological design reduces climate risk by creating production systems that are well adapted to the constraints of local landscape and climate conditions, enhancing the production of ecosystem services that buffer production from weather-related disturbance. High-value markets, for example direct and specialty markets, reduce climate risk through improved profitability and the production of social capital, which enhance the capacity to respond to challenging climate conditions and recover from climate-related damages.

Adaptive management has most commonly been used by natural resource managers for the sustainable development of fisheries, forests and native rangelands and in biodiversity conservation and ecosystem restoration efforts. More recently, it has been applied to sustainable business and community development, including sustainable agriculture. Adaptive management principles emphasize active learning and the discovery of robust solutions to challenges through a continuous process of monitoring, evaluating and adjusting management actions based on what is learned through monitoring system behavior.[24] Adaptive management has proven useful in the management of complex systems characterized by high uncertainty and dynamic change.

Resilience thinking offers a conceptual framework particularly useful to the management of agricultural production systems under the challenges of resource scarcity and climate change. It extends sustainable agriculture management considerations with the explicit recognition that agricultural production systems, like all social and ecological systems, exist in a state of dynamic change. In addition, resilience thinking acknowledges that social and ecological systems do not operate independently but rather interact across different spatial scales and through time in highly complex and sometimes surprising ways.[25] Applied to the design and management of sustainable agriculture systems, resilience thinking practices will enhance adaptive capacity to climate change effects while furthering economic, social and environmental sustainability goals.[26] Agricultural resilience is being actively explored by researchers working at a variety of scales, including on the farm,[27] in rural communities[28] and across regions.[29]

Although the potential for adaptive management and resilience thinking to improve climate risk management has yet to be fully developed, the principles and practices of sustainable agriculture represent a well-developed framework to enhance the adaptive capacity of US agriculture to climate change and resource scarcity.[30]

Enhancing Adaptive Capacity with Sustainable Agriculture

Sustainable farmers and ranchers throughout the country manage a wide range of diversified production systems that are productive and profitable businesses. Each production system is a unique expression of local productive capacity and the creativity and style of the manager. Many sustainable farms and ranches have supported multiple generations through the dynamic changes that accompanied the industrial transformation of American agriculture. Indeed, many producers began making the transition to sustainable practices in response to the social and environmental challenges created by the industrialization of the US food system.

Sustainable producers promote social and economic well-being by managing for ecological health—both on the farm or ranch and in their communities—with practices that aim to enhance the natural resource base upon which agriculture depends while achieving the economic and social goals of sustainability. Although the environmental and productivity benefits of sustainable agriculture practices have been well-documented by agricultural researchers, the economic, social justice and community benefits have received less attention.[31] Research exploring economic and social performance of sustainable agriculture is complex and difficult to interpret because the operating context has a large influence on production system performance; however, a large body of case study research documents the economic and social benefits of sustainable agriculture throughout the United States.[32]

Economic Sustainability: Managing for Profitability

Economic sustainability involves managing the agricultural business to ensure the financial well-being over the long term of everyone involved in it as owners, managers or laborers. For an agricultural business to achieve the highest level of economic sustainability, total earnings need to cover all production and management costs, provide living wages and benefits to management and employees, with some left over each year to provide capital for investment and the recovery needs of the business over the long term.[33]

Practices that promote economic sustainability include managing enterprises for profitability, sound financial planning, proactive marketing, effective risk management and good overall management.[34] While many sustainable practices contribute to economic sustainability, the careful design and management of profitable diversified crop and livestock enterprises suited to the unique environment of the production system, high-value and value-added products, innovative marketing strategies, and values-based food supply chains will likely become increasingly important climate risk management tools. Diversification spreads financial risk across many enterprises, and the increased profitability possible with high-value products and markets improves financial viability and recovery reserves.

Social Sustainability: Managing for Family and Community Well-being

Social sustainability means maintaining a high quality of life for the people that live and work on the farm or ranch and in the community over the long term.[35] It includes practices such as maintaining good communication, trust and mutual support. Practices that promote a high quality of life on the farm or ranch include full participation in planning, creating space to talk openly and honestly, making progress toward production system goals and having a general feeling of satisfaction and happiness. Social sustainability recognizes the role that human resources play in production system performance. Sustainable practices strive to encourage everyone involved in the farm or ranch to contribute their ideas and effort toward mutually agreed upon goals.[36] Providing opportunities for everyone on the farm or ranch to develop and grow and take on new challenges and responsibilities invites full participation in managing for success and improves quality of life for farm or ranch residents, managers and employees.

Social sustainability extends to the community within which the production system is located. There are many sustainable practices that promote community well-being, including hiring seasonal labor from the local community, supporting local businesses, connecting with local consumers and inviting them on the farm or ranch, taking a consumer-oriented approach to production and management practices and teaching the community about sustainable food production. Marketing strategies such as community supported agriculture, farmers' markets and other kinds of local and regional marketing, hosting school tours and summer camps, and offering farm internships all enhance the sustainability of the farm by increasing knowledge and awareness of food production and cultivating mutual understanding and support between food producers and consumers.

Practices that enhance social sustainability on farms and ranches and in communities increase the adaptive capacity of the production system to climate change effects. The development of human resources on the farm or ranch may improve production system performance, particularly under conditions of uncertainty and dynamic change, and is likely to enhance the capacity for innovation.[37] The social and financial capital generated by the development of community sustainability will increase the resources available to enhance adaptive capacity and maintain recovery reserves.

Ecological Sustainability: Managing for Ecosystem Health

Sustainable producers manage extremely complex systems characterized by many dynamic linkages in the production system and with systems beyond the farm or ranch gate. Agricultural production systems are a special kind of ecosystem designed and managed by humans for specific social purposes. Like an ecosystem, an agricultural production system can be understood through the study of its structural and functional properties—for example, how energy, water and nutrients enter, move through and exit the system.

4.3. Agroecology: The Science of Agricultural Ecosystems

Agroecology is the study of the ecological processes that operate in agriculture and food systems.[38] An interdisciplinary science, agroecology draws on many biological, agricultural and social science disciplines to understand the ecology of agroecosystems and develop the knowledge needed to enhance the sustainability of agricultural and food systems. The environmental, social and economic benefits of sustainable agriculture emerge from agroecosystems that are designed and managed to promote healthy ecosystems.

An ecosystem is a community of living organisms interacting with each other and the non-living environment within which the organisms live so that energy flows and materials cycle. The ecosystem is fundamental to the understanding of life on Earth and provides the biological basis for ecological sustainability and resilience as emergent properties of complex adaptive systems.

Complex adaptive systems are systems that have independent and interacting components subject to selection pressures that produce constant variation and novelty through changes in existing components or the entry of new components to the system. An emergent property of a system is a characteristic or behavior that arises from the interactions between the component parts of the system and cannot be observed in the components in isolation from the system. For example, health, life and resilience are emergent properties of a biological organism, and yield, soil quality, profit, sustainability and resilience are emergent properties of a farm or ranch system.

Ecosystems can be described in terms of some key structural and functional properties that largely determine their health and productivity. The structural properties—such as species diversity, vegetative architecture and the food web—describe the physical relationships between the organisms that inhabit the ecosystem. Functional properties describe the dynamic processes that capture, move and store energy and materials in the ecosystem, regulate populations of organisms that inhabit it and shape the development of the system over time. These are processes like photosynthesis, herbivory and predation, and soil decomposition. Although designed and managed by humans for the purpose of producing harvestable materials, the basic structural and functional processes that occur in all natural ecosystems can be observed and managed in agroecosystems to promote sustainability and resilience.

Ecosystems are defined by an arbitrary boundary that is selected by the observer. Energy flow and material cycling can be explored between components of a field, a whole farm or ranch or even an entire agricultural region. In each case, the physical components of the ecosystem can be defined and measured, the interactions between components investigated and emergent properties like crop and livestock health, profitability, sustainability and resilience explored.

Farmers, growers and ranchers are ecosystem engineers—they design and build the farm system and are the conductors of a dynamic symphony of ecological processes that ultimately produce marketable goods and community services. To comprehend and manage this dynamic complexity, sustainable producers focus on four fundamental ecological processes—energy flow, water and nutrient cycling and community dynamics—that operate in all ecosystems, including agroecosystems.[39] These four processes, essential to the ecosystem services provided to agriculture, are linked to each other in complex ways. A change to the functioning of one process will likely change the functioning of the other three, and the healthy functioning of each process is required to support the healthy function of the others.

Take nitrogen, for example. In order to supply nitrogen, an essential plant nutrient required in large amounts by many crops, the agroecosystem must have sufficient *community dynamics* to capture the element from the atmosphere, store it in plant tissues and release it for plant use. Each of these processes requires the activities of a diversity of organisms that include plants, animals and microorganisms. All organisms in the agroecosystem require a sufficient flow of *energy, water* and *nutrients* to capture, store and release nitrogen. All four processes work together to support ecosystem health, which is the foundation of agroecological sustainability.

These four ecosystem processes are easy to observe in agricultural production systems and can be managed quite effectively with currently available knowledge and technologies. The regular observation of these four ecosystem processes offers producers a useful way to assess the effects of management decisions, as well as conditions they cannot manage, such as climatic effects, on the overall health and productivity of the agroecosystem.

Energy Flow

Agriculture is in the business of harvesting sunlight. Vegetable and fruit growers, cattle ranchers and grain farmers, whether using industrial or sustainable practices, are all managing plants to capture sunlight and produce commercial products, directly or indirectly (through the production of livestock). Managing the agroecosystem to maximize the capture, storage and efficient use of solar energy is a basic sustainable agriculture strategy.

Solar energy capture can be enhanced by design and management of the agricultural landscape to maximize, in space and time, the amount of actively photosynthesizing leaf area. Diverse crop rotations and cover crops, management-intensive grazing, season extension practices and the addition of perennial plants as crops or landscape features are key tools that sustainable producers use to capture more solar energy with the agricultural production system.

As fossil fuel energy supplies become more limited and efforts to mitigate global warming expand, more effective use of solar energy management strategies will become increasingly important to the adaptive capacity of US agriculture. Sustainable producers have been busy over the last forty years working on the design and improvement of an astonishing number of diversified production systems that effectively maximize the capture and flow of solar energy in agroecosystems, make efficient use of non-renewable energy and even produce energy on the farm or ranch while supporting the multiple goals of sustainable agriculture.[40]

The Water Cycle

The availability of water has been identified as one of the most critical natural resource challenges facing US agriculture in the twenty-first century.[41] Agriculture is a major user of water in the United States, accounting for about 80 percent of consumptive water use nationally. More than 75 percent of all irrigated cropland and pasture land is found in the western states, where agriculture accounts for over 90 percent of all consumptive water use.[42]

The unsustainable use of freshwater resources, combined with the degradation of water quality through the release of agricultural, industrial and domestic wastes, have reduced the quantity and quality of freshwater supplies available to agriculture and other users. Climate change will create additional challenges to water resources because higher temperatures and drought will increase water demand, and more variable precipitation and more heavy rainfall will increase soil erosion and runoff and reduce groundwater recharge.[43]

Sustainable producers manage soil quality to promote a healthy water cycle on their farm or ranch. High-quality soils maximize the capture and storage of precipitation entering the agroecosystem by promoting rapid surface infiltration with a strong, porous soil surface structure and a high soil-water storage capacity. Practices that produce high-quality soils include the use of cover crops and crop rotation to build soil organic matter and protect the soil surface from erosion by wind or water. Where topography is less hospitable to annual cropping systems, perennial cropping systems such as grass-based animal systems or high-value fruit and nut production systems can be managed to produce high soil quality.

Sustainable producers use a number of common practices to use water efficiently on their farms and ranches. Crop rotations are designed to fit seasonal temperature and precipitation patterns in order to reduce or eliminate the need for supplementary irrigation. When irrigation is needed, sustainable producers are more likely to use efficient application systems,[44] and high soil quality reduces overall irrigation needs because soil water storage is high and runoff losses are minimized. Some producers capture and manage surface water storage in lakes, ponds, riparian areas and seasonal wetlands.[45]

The Nutrient Cycle

Along with water and temperature, nutrient availability is an important factor limiting plant growth. Plants require eighteen essential elements, also called plant nutrients, to grow and develop normally. All but two of these elements—carbon and oxygen—are supplied to plants through soil processes. Carbon is the dominant essential element in all life on Earth and typically makes up about half of total plant mass in a land-based ecosystem.

Carbon cycles between the atmosphere, plants and soil through photosynthesis, cellular respiration and soil decomposition processes. Photosynthesis captures carbon from the atmosphere and stores it in plant tissues. Respiration and decomposition release carbon dioxide back into the atmosphere. Carbon compounds commonly include large amounts of oxygen, taken up from the atmosphere, and hydrogen, from soil water, as well as smaller amounts of the other essential elements. The cycling of the essential elements is governed, in large part, by the cycling of carbon through the ecosystem.

In natural ecosystems, nutrients tend to be conserved in soil organic matter and plant and animal tissues. The nutrients are used over and over again by different organisms as they flow through a cycle that includes plant and animal growth—which takes essential nutrients from the soil, uses them in growth and maintenance and returns them to the soil in waste products and through death and decomposition. The nutrients in plant and animal wastes cannot be reused until soil decomposition processes release them in a form available to plants. In natural systems, this cycling of nutrients between plants, animals and the soil is extremely conservative and few nutrients are lost; however, the very purpose of agroecosystems—to produce harvested materials—makes nutrient conservation more challenging, and essential nutrients are lost from the system through a number of pathways, both intentional and unintentional.

Nutrients are intentionally removed by the harvest and sale of plant and animal products and are also lost in soil drainage waters (mostly nitrogen), in gasses produced by soil decomposition processes (mostly nitrogen) and in surface runoff waters (many nutrients, plus soil organic matter and pesticides). The unintentional loss of these nutrients from the agroecosystem reduces productivity and increases the costs of production. These losses are also a major factor in the pollution of ground and surface waters in the United States and include heat-trapping gases that contribute to global warming.

The aim of sustainable management of a healthy nutrient cycle is to maximize the cycling of nutrients in the agroecosystem—to provide growing crops the nutrients they need, when they need them—while minimizing unintentional losses to the environment. A healthy nutrient cycle supports high yields of healthy crops and livestock, reduces the need to purchase plant nutrients and livestock feeds, and contributes to environmental quality through carbon sequestration and reductions in nutrient enrichment of surface and groundwater and emissions of heat-trapping gases to the atmosphere.

Sustainable agriculture practices that promote a healthy nutrient cycle include the integration of crops and livestock, feeding livestock with crops produced in the production system, careful management of animal manure and crop residues in the field and through composting, managing for balance between macronutrients entering and leaving the production system, the use of catch crops and diverse crop rotations to reduce nutrient leaching losses and practices that prevent soil erosion by wind and water.[46]

Community Dynamics

Community dynamics—the interactions between species in the agroecosystem—is the fourth ecosystem process actively managed by sustainable producers. In agroecosystems, community dynamics are planned and managed by the producer to enhance functional biodiversity on the farm or ranch. Energy flow, water and nutrient cycles and soil quality all depend on a diversity of species needed to carry out soil decomposition processes, nutrient transformations, pest suppression, pollination and other ecological processes responsible for providing ecosystem services to agriculture.

Community dynamics are enhanced not just by managing for a large number of species, but through diversity that has been planned to support agroecosystem structure and function. For example, a common sustainable practice on farms that raise and finish cattle and hogs is to foster a healthy nutrient cycle through the careful selection of crops grown in rotation. Soils produce grain crops to feed the hogs and forage crops to feed the cattle and all four elements of the farm design—grain, forage, livestock, soil—are linked through the nutrient cycle. The forage crops are carefully selected to accumulate crop nutrients and build soil quality. After about three years in forages, the field is planted in grain crops, which benefit from the improved soil quality and use the nutrients released through the decomposition of forage crop residues. Livestock are often moved through the forage and grain fields at specific points in the crop rotation to return plant nutrients to the field in their manure for storage and recycling. This sustainable design builds soil quality and conserves nutrients in the agroecosystem through the careful design of linkages that encourage the cycling of nutrients between crops, livestock and soil.

Key sustainable practices that enhance community dynamics include crop rotation and intercropping, the integration of annual and perennial crops, cover cropping and the use of insectary crops, the integration of crops and livestock, the production of multiple livestock species, multiple types and different age groups within livestock species and the intentional management of field edges, fence rows and riparian areas to promote biodiversity. These planned biodiversity practices are proven to reduce the costs of purchased inputs, improve soil quality, enhance nutrient and water conservation, pollinate crops, suppress pest populations and contribute to improved environmental quality in the community and beyond. Biodiversity at the field, farm, ranch and landscape scale is a key feature of a productive, robust and resilient agriculture.[47]

Putting It All Together: Systems of Sustainable Agriculture

Contemporary agriculture systems that achieve ecological, social and economic sustainability goals can be found throughout the United States. These systems are designed and managed by farmers and ranchers who work to enhance farm assets—human, natural, social, physical and financial—through the use of ecological design principles that can be identified with one of five strategies of ecosystem-based agriculture: biointensive, permaculture, biodynamic, organic and low-input. All five systems place a high value on soil quality and biodiversity, but each is distinguished through unique principles and associated soil, crop and livestock management practices. The many diverse expressions of these ecologically based food production systems offer a wealth of innovative strategies that could enhance the adaptive capacity of US agriculture and food systems.

Biointensive agriculture is a small-scale, labor-intensive system of food production that places an emphasis on intensive management to produce extremely high crop yields through the use of deep soil preparation, composting and intensive planting. These practices, managed in a whole system that includes carbon and calorie crops and seed saving, result in reductions in water, fertilizer and fossil fuel inputs of 50 to 90 percent and a two- to four-fold increase in calorie production per unit area over industrial production systems. The intensive focus on deep soil preparation and on-farm compost production, combined with resource use reductions, offer significant advantages to both climate change mitigation and adaptation.[48] Ecology Action, a non-profit organization based in California, conducts research, provides educational programs and supports the development of biointensive agriculture throughout the world.[49]

Permaculture is an ecological community design system that includes an emphasis on food production in polycultures of annual and perennial plants. Although more typically applied to small-scale subsistence food production in tropical and subtropical regions, interest in the potential of permaculture designs for large-scale, commercial agricultural systems in temperature regions is growing. New Forest Farm, located in southwest Wisconsin, offers one well-known example of a temperate, large-scale perennial permaculture production system modeled on the native oak savannah biome.[50] Research at the Land Institute in Kansas has focused for more than thirty years on the development of perennial grain production systems modeled on the native prairie biome.[51] The Savannah Institute is leading a new research effort to document the costs and benefits of large-scale permaculture production systems on farms in the Midwest.[52] The use of perennial plants, particularly trees, and the resource use reductions possible in polyculture production systems offer advantages to both climate change mitigation and adaptation.[53]

Biodynamic agriculture is an integrated crop and livestock production system that places an emphasis on managing on-farm resources for soil, crop and livestock health and vitality through the use of crop diversity, composts, microbial inoculants and herbal remedies.[54] More commonly used on commercial farms in Europe, biodynamic practices are less well-known in North America; however, interest among US wine producers, impressed by the quality improvements possible with biodynamic practices, has increased awareness among farmers and consumers in the United States. Biodynamic producers benefit from an international certification program, managed by the Demeter Association, which provides guidelines for soil, crop and livestock management and identifies products generated using biodynamic practices as value-added in commercial markets.[55] Fred Kirschenmann's 2,600-acre grain and livestock farm in North Dakota offers a model of a mid-sized biodynamic farm.[56] A focus on the production of high-quality composts and the use of on-farm resources offer advantages to both climate change mitigation and adaptation.

Organic agriculture is an ecological design system that places an emphasis on the production of healthy crops and livestock through practices that enhance soil health and environmental quality. Widely recognized by agriculturalists and consumers worldwide and practiced at scales ranging from small home gardens to large industrial production systems, commercial organic producers benefit from national and international certification programs such as USDA's National Organic Program, which provide guidelines for soil, crop and livestock management and identifies products produced with organic practices as value-added in commercial markets. Many of the farms featured in this book offer models of sustainable certified organic production, for example, vegetables (Paul Muller in California), fruits (Jim Koan in Michigan), grains (Bob Quinn in Montana) and livestock products (Will Harris in Georgia). The focus on soil health and environmental quality offers advantages to both climate change mitigation and adaptation. A recent analysis suggests regenerative organic agriculture has the capacity to sequester more than 100 percent of current global carbon dioxide emissions if adopted worldwide.[57]

Low-input agriculture is a farming system design that blends industrial and organic practices to promote soil quality and healthy ecosystem processes and reduce the need for off-farm inputs like fertilizers and pesticides.[58] Low-input farmers use more diversified crop rotations, integrate livestock into crop production systems, practice management-intensive grazing and use integrated pest management, but they also may use some synthetic fertilizers and pesticides when needed. Low-input agriculture aims to reduce the impact of farm practices on the environment while maintaining productivity and increasing profitability. Many of the producers featured in this book offer models of low-input farming systems producing potatoes (Brendon Rockey), nuts (Dan Shepherd), fruit and vegetables (Jonathan Bishop), grains (Russ Zenner, Gabe Brown, Gail Fuller) and livestock (Gary Price, Julia Davis Stafford, Mark Frasier). The focus on soil health and the reduction of energy-intensive fertilizers and pesticides offer advantages to both climate change mitigation and adaptation.

Lessons from the Field: Is Sustainable Agriculture Climate Ready?

Sustainable agriculture practices are proven to promote the ecological, economic and social well-being of farmers and society; however, there is little direct evidence that these practices reduce climate risks to US farms and ranches. Do crop diversity, integrated crop and livestock production, and soil quality buffer climate variability and extremes? Does the increased emphasis on building natural, social and human assets on farms and in communities improve the performance of sustainable farms and ranches in more variable and extreme weather conditions? Can the adaptive capacity of industrial agriculture be enhanced through the use of sustainable practices? What are the barriers and opportunities to the development of a sustainable US agriculture robust to the increasing pace and intensity of climate change? In the next four chapters, twenty-five award-winning sustainable farmers and ranchers with long experience managing the production of vegetables, fruits and nuts, grains and livestock across the United States share their answers to these questions.

Endnotes

1. Science Daily. 2009. Deadly Heat Waves Are Becoming More Frequent in California. sciencedaily.com/releases/2009/08/090825151008.htm
2. A. Nossiter. 2007. "Drought Saps the Southeast, and Its Farmers." *New York Times*. August 4th nytimes.com/2007/07/04/us/04drought.html
3. M. Johnson. 2008. Floods Create Economic Catastrophe in Midwest. NBC News. nbcnews.com/id/25274045/ns/business-stocks_and_economy/t/floods-create-economic-catastrophe-midwest/#.U3EfPyjPZoE
4. S. Melker. 2012. A Sour Season for Michigan's Cherry Farmers. PBS News Hour pbs.org/newshour/updates/science-july-dec12-michigancherry_08-15/
5. It's Official: 2012 Drought Cost Taxpayers a Record 14 Billion. 2013. Taxpayers for Common Sense. taxpayer.net/library/article/2012-drought-cost-taxpayers-a-record-14-billion
6. T. Waters. 2013. U.S. Drought in 2013 Hurts Cattle Ranchers With Dry, Poor Wheat Crop. Huffington Post. huffingtonpost.com/2013/01/14/us-drought-2013-cattle-ranchers_n_2469742.html
7. J. Hurdle. 2014. Climate Check: US Produce Prices to Rise on Extreme CA Drought. MNI Market News. mninews.marketnews.com/content/climate-check-us-produce-prices-rise-extreme-ca-drought-0
8. See chapters 2 and 3 for a detailed discussion of role of exposure and sensitivity in determining the potential impact of climate change on a farm or ranch.
9. N. Marshall et al. 2009. Enhancing Adaptive Capacity. Ch. 15 in C. Stokes and M. Howden (ed.), *Adapting Agriculture to Climate Change: Preparing Australian Agriculture, Forestry and Fisheries for the Future*. CSIRO Publishing: Collingwood, Victoria, Australia.
10. Ibid.

11. Adapted from Table 15.1, Ibid.
12. C. Walthall et al. Agricultural Water Resources and Irrigation, Climate Change Effects on U.S. Agricultural Production. Ch. 5 in *Climate Change and Agriculture in the United States: Effects and Adaptation*. 2012. USDA Technical Bulletin 1935.
13. C. Walthall et al. Ecosystem Services, Climate Change Effects on U.S. Agricultural Production. Ch. 5 in *Climate Change and Agriculture in the United States: Effects and Adaptation*. 2012. USDA Technical Bulletin 1935.
14. C. Walthall et al. Agricultural Soil Resources, Climate Change Effects on U.S. Agricultural Production. Ch. 5 in *Climate Change and Agriculture in the United States: Effects and Adaptation*. 2012. USDA Technical Bulletin 1935.
15. National Research Council. Drivers and Constraints Affecting the Transition to Sustainable Farming Practices. Ch. 6 in *Toward Sustainable Agricultural Systems in the 21st Century*. 2010. The National Academies Press, Washington, DC.
16. N. Marshall et al. 2009. Enhancing Adaptive Capacity. Ch. 15 in C. Stokes and M. Howden (ed.), *Adapting Agriculture to Climate Change : Preparing Australian Agriculture, Forestry and Fisheries for the Future*. CSIRO Publishing: Collingwood, Victoria, Australia.
17. C. Walthall et al. Social Barriers to Adaptation, Adapting to Climate Change. Ch. 7 in *Climate Change and Agriculture in the United States: Effects and Adaptation*. 2012. USDA Technical Bulletin 1935.
18. P. Howe and A. Leiserowitz. 2013. Who remembers a hot summer or a cold winter? The asymmetric effect of beliefs about global warming on perceptions of local climate conditions in the U.S. *Global Environmental Change*, 23: 1488–1500.
19. Results of Midwest and Northern Great Plains survey reported in J. Arbuckle Jr. et al. 2013. Climate Change Beliefs, Concerns, and Attitudes toward Adaptation and Mitigation among Farmers in the Midwestern United States. Climatic Change Letters 117:943–950. Results of MS, TX, NC and WI survey reported in R. Rejesus et al. 2013. U.S. Agricultural Producer Perceptions of Climate Change. *Journal of Agricultural and Applied Economics*, 45:701–18.
20. J. Jemison Jr. et al. 2014. How to Communicate with Farmers About Climate Change: Farmers' Perceptions and Adaptations to Increasingly Variable Weather Patterns in Maine. *Journal of Agriculture, Food Systems and Community Development* 4(4): 57-70.
21. C. Walthall et al. Managing Climate Risk: New Strategies for Novel Uncertainty, Adapting to Climate Change. Ch. 7 in *Climate Change and Agriculture in the United States: Effects and Adaptation*. 2012. USDA Technical Bulletin 1935.

22. National Research Council. Understanding Agricultural Sustainability. Ch. 1 in *Toward Sustainable Agricultural Systems in the 21st Century*. 2010. National Academies Press, Washington, DC.
23. For example, the Practical Farmers of Iowa, Ecological Farming Association, National Sustainable Agriculture Coalition, USDA's Sustainable Agriculture Research and Education program, and many regional and state sustainable agriculture organizations such as the Northeast Organic Farming Association, the Carolina Farm Stewardship Association, Delta Land and Community, Kerr Center for Sustainable Agriculture, Northern Plains Sustainable Agriculture Society, the National Center for Appropriate Technology, and Oregon Tilth.
24. K. Moore. 2009. Landscape Systems Framework for Adaptive Management. Ch. 1 in *The Sciences and Art of Adaptive Management: Innovating for Sustainable Agriculture and Natural Resource Management*. Ankeny, IA: Soil and Water Conservation Society.
25. B. Walker and D. Salt. 2006. Living in a Complex World: An Introduction to Resilience Thinking. Ch. 1 in *Resilience Thinking: Sustaining Ecosystems and People in a Changing World*. Island Press: Washington DC.
26. National Research Council. Understanding Agricultural Sustainability, Ch. 1 in *Toward Sustainable Agricultural Systems in the 21st Century*. 2010. National Academies Press: Washington, DC.
27. J. Hendrickson et al. 2008. Interactions in integrated US agricultural systems: The past, present and future. *Renewable Agriculture and Food Systems*, 23(04): 314–324; P. Reidsma et al. 2010. Adaptation to climate change and climate variability in European agriculture: The importance of farm level responses. *European Journal of Agronomy*, 32(1): 91–102; D. F. van Apeldoorn et al. 2011. Panarchy rules: Rethinking resilience of agroecosystems, evidence from Dutch dairy-farming. *Ecology and Society*, 16(1); N. A. Marshall. 2010. Understanding social resilience to climate variability in primary enterprises and industries. *Global Environmental Change*, 20(1): 36–43; I. Darnhofer. 2009. Strategies of family farms to strengthen their resilience. 8th International Conference of the European Society for Ecological Economics; I. Darnhofer et al. 2010. Adaptiveness to enhance the sustainability of farming systems: A review. *Agronomy for Sustainable Development*, 30 (3):545–555; I. Darnofer et al. 2010. Assessing a farm's sustainability: Insights from resilience thinking. *International Journal of Agricultural Sustainability*, 8(3): 186–198.
28. R. Nelson et al. 2010. The vulnerability of Australian rural communities to climate variability and change: Part I—Conceptualising and measuring vulnerability. *Environmental Science &Policy*, 13(1): 8–17; R. Nelson et al. 2010. The vulnerability of Australian rural communities to climate variability

and change: Part II—Integrating impacts with adaptive capacity. *Environmental Science & Policy*, 13(1): 18–27; J. Arbuckle. 2011. Farmer perspectives on climate change and agriculture. Climate Change Conference. Heartland Regional Water Coordination Initiative, Lied Conference Center, Nebraska City, NE.

29. H. Allison and R. Hobbs. 2004. Resilience, adaptive capacity, and the "lock-in trap" of the western Australian agricultural region. *Ecology and Society*, 9(1): 3; D. Wolfe et al. 2008. Projected change in climate thresholds in the northeastern U.S.: Implications for crops, pests, livestock, and farmers. *Mitigation and Adaptation Strategies for Global Change*, 13(5): 555–575; W. Easterling. 2009. Guidelines for adapting agriculture to climate change, in D. Hillel and C. Rosenzweig (ed.), *Handbook of Climate Change and Agroecosystems : Impacts, Adaptation, and Mitigation*. Imperial College Press; L. Jackson et al. 2011. Case study on potential agricultural responses to climate change in a California landscape. *Climatic Change*, 109(Suppl 1): S407–S427; G. Kenny. 2011. Adaptation in agriculture: Lessons for resilience from eastern regions of New Zealand. *Climatic Change*, 106(3): 441–462.
30. National Research Council. Understanding Agricultural Sustainability. Ch. 1 in *Toward Sustainable Agricultural Systems in the 21st Century*. 2010. National Academies Press: Washington, DC.
31. National Research Council. Economic and Social Dimensions of the Sustainability of Farming Practices and Approaches. Ch. 4 in *Toward Sustainable Agricultural Systems in the 21st Century*. 2010. National Academies Press: Washington, DC
32. For example, see *The New American Farmer: Profiles of Agricultural Innovation*, Second Edition. 2005. Sustainable Agriculture Network. USDA-SARE and National Research Council. Illustrative Case Studies, Ch. 7 in *Toward Sustainable Agricultural Systems in the 21st Century*. 2010. National Academies Press: Washington, DC.
33. L. Lengnick and S. Kask. 2009. Final Report of the Modelling Team, Prosperity Project: Improved Farm Profit with Farmland Protection and High Value Crops Program, NRI Grant # 2005-35618-15645. Submitted to the National Research Initiative of the Cooperative State Research, Education and Extension Service, USDA.
34. P. Sullivan. 2003. Applying the Principles of Sustainable Farming. National Sustainable Agriculture Information Service, National Center for Appropriate Technology: Butte, MT.
35. Ibid.
36. E. Henderson and K. North. 2011. Whole Farm Planning: Ecological Imperatives, Personal Values and Economics, Second Edition. Northeast Organic

Farming Association (NOFA) Interstate Council, Organic Principles and Practices Handbook Series.

37. C. Walthall et al. Managing Climate Risk: New Strategies for Novel Uncertainty, Adapting to Climate Change. Ch. 7 in *Climate Change and Agriculture in the United States: Effects and Adaptation*. 2012. USDA Technical Bulletin 1935.
38. Adapted from S. Gliessman. 2007. The Agroecosystem Concept, Ch. 2 and Agroecosystem Diversity and Stability. Ch. 16 in *Agroecology: The Ecology of Sustainable Food Systems*. Second Edition. CRC Press: New York.
39. P. Sullivan. 2003. Applying the Principles of Sustainable Farming. National Sustainable Agriculture Information Service, National Center for Appropriate Technology: Butte, MT.
40. For example, see *Clean Energy Farming: Cutting Costs, Improving Efficiencies, Harnessing Renewables*. 2008. USDA Sustainable Agriculture Research and Education Program: Washington, DC.
41. National Research Council. A Pivotal Time in U.S. Agriculture. Ch. 2 in *Toward Sustainable Agricultural Systems in the 21st Century. 2010*. National Academies Press: Washington, DC.
42. G. Schaible and M. Aillery. 2012. Water Conservation in Irrigated Agriculture: Trends and Challenges in the Face of Emerging Demands. USDA Economic Information Bulletin 99.
43. C. Walthall et al. Agricultural Soil Resources: Climate Change Effects on U.S. Agricultural Production. Ch. 5 in *Climate Change and Agriculture in the United States: Effects and Adaptation*. 2012. USDA Technical Bulletin 1935.
44. For examples of sustainable water management practices on U.S. farms and ranches see, *Smart Water Use on Your Farm or Ranch*. 2006. USDA SARE. Sustainable Agriculture Network: Washington, DC.
45. F. Magdoff and H. van Es. 2009. Managing for High Quality Soils. Ch. 9 in *Building Soils for Better Crops: Sustainable Soil Management*, Third Edition. Handbook Series Book 10. USDA SARE: Washington, DC.
46. F. Magdoff and H. van Es. 2009. Nutrient Management: An Introduction. Ch. 18 in *Building Soils for Better Crops: Sustainable Soil Management*, Third Edition. Handbook Series Book 10. USDA SARE: Washington, DC.
47. National Research Council. Improving Productivity and Environmental Sustainability. Ch. 3 in *Toward Sustainable Agricultural Systems in the 21st Century*. 2010. National Academies Press, Washington, DC.
48. J. Beebe. 2010. Climate Change and Grow Biointensive. Ecology Action.
49. Ecology Action: Grow Biointensive. growbiointensive.org
50. M. Sheppard. nd. New Forest Farm. newforestfarm.net
51. W. Jackson. nd. The Land Institute. landinstitute.org
52. The Savanna Institute is a non-profit organization dedicated to developing

restorative, savanna-based agricultural systems through research, education and outreach. savannainstitute.org

53. Permaculture Principles. permacultureprinciples.com
54. The Biodynamic Association is a non-profit association of individuals, groups and organizations who are committed to rethinking agriculture through healthy food, healthy soil and healthy farms. Founded in 1938, the BDA is considered to be the oldest sustainable agriculture organization in North America. biodynamics.com
55. The Demeter Association is a non-profit organization with the mission to enable people to farm successfully, in accordance with Biodynamic practices and principles. demeter-usa.org
56. P. Pearsall. 2013. Farmer-Philosopher Fred Kirschenmann on Food and the Warming Future. *Yes Magazine*. yesmagazine.org/planet/farmer-philosopher-fred-kirschenmann-farms-warming-future
57. Regenerative Organic Agriculture and Climate Change: A Down-to-Earth Solution to Global Warming. 2014. Rodale Institute. rodaleinstitute.org/assets/WhitePaper.pdf
58. D. Pimentel et al. 1989. Low-input sustainable agriculture using ecological management practices. *Agriculture, Ecosystems & Environment*, Vol. 27 (1): 3–24.

CHAPTER 5

Vegetables

The sustainable vegetable growers featured in this book are located in nearly every major region of the country. They represent a variety of models of successful diversified vegetable production, ranging in scale from four to five hundred acres, and have experience managing vegetable production at one location ranging from twenty to more than forty years. Many use certified organic practices, and all view soil quality, biodiversity and orientation with value-added direct and direct-wholesale markets as key production risk management strategies. All of these award-winning vegetable growers report changes in weather conditions over the last decade or so that are outside their experience of normal variability or extremes.

Vegetable growers in the West have been challenged by more variable temperatures and precipitation, but have also found opportunity in the longer growing season. Water conservation has been and will continue to be a high priority in this arid region, and securing a supply of water of sufficient quantity and quality for vegetable production is likely to become more challenging as climate change intensifies. Western growers have grave concerns about their continued access to water as warmer winters, drought and higher summer temperatures reduce water supplies and increase competition for water in the region.

Eastern growers are challenged by more temperature and precipitation extremes, more intense heat waves, excessive rainfall, warmer springs, higher humidity and a startling increase in crop damage from novel plant diseases. More heavy precipitation and longer drought periods have complicated the timely completion of fieldwork such as soil preparation, planting and cultivation, because soils are often too wet or too dry to work without damage. Extreme swings in temperature, particularly in winter and spring, and high temperatures in mid-summer have disturbed regular crop development and reduced crop quality and yields. Traditional diseases are appearing earlier and are more damaging, perhaps because of higher humidity and earlier spring warm-up, while novel diseases have proven particularly damaging.

The Northwest

Nash Huber, Nash's Organic Produce, Sequim, Washington

Nash Huber has been farming in the north of Washington's Olympic Peninsula for more than forty years. Over that time, Nash has assembled a productive and profitable organic farm that produces an incredible diversity of foods marketed year-round from the farm and to wholesale and direct markets in the Seattle region. Together with a permanent crew of twenty-five that grows to forty during the peak of the growing season, Nash produces organic vegetables and fruits, food and feed grains, pork and poultry and a variety of vegetable and cover crop seeds on 450 acres of prime farmland, much of it leased and most of it protected by conservation easements.

FIGURE 5.1. Nash Huber and Patty McManus-Huber, Nash's Organic Produce, Sequim, Washington. Credit: Linda Barnfather

Nash uses diverse rotations, cover crops, insectary plantings and farm-made compost to create the biodiversity and soil quality needed to provide nutrients, conserve water and reduce pest pressures and keep the farm productive and profitable.

The farm receives an annual average of seventeen inches of rain, and all its production is irrigated with surface waters replenished each year by snowmelt from the mountains to the south.

Nash Huber first started noticing changes in the weather in the early 1990s. The growing season seemed to lengthen as winters grew warmer, spring temperatures and precipitation grew more variable and fall grain and seed harvests were increasingly disturbed by more frequent fall moisture. "It seems like our springs have gotten longer, cooler and wetter. We always used to get nice warm weather in late April and May. We haven't gotten anything like that in years. Quite often, spring will start sometime between the middle of January and the middle of March, and it's really variable. It can have too much swing, and that really puts the squeeze on us."

> Winters are warmer, falls are wetter, and summers are more unpredictable. The challenge is just the instability of not being able to know where we are headed.
>
> — Nash Huber

Wetter springs complicate the management of spring crops for a number of reasons. Wide swings in the timing of the spring warm-up make it difficult to plan a spring planting schedule, and wet conditions increase the difficulty of killing cover crops and preparing soils for spring planting in a timely manner as well as complicating weed management. "The weeds love to come on in spring time. If you can't get into the field and cultivate in a timely manner, you got increased weed pressure." Cool, wet conditions also increase disease pressures in spring crops, slow crop growth and lengthen time to harvest.

Highly variable summers and increased moisture in the fall have a big impact on the farm's most valuable products—grains and seeds. September weather used to be predictable, Nash explains: "You could count on thirty days of clear, sunny weather, but it is no longer that way. It used to be that our marine fog didn't start to show itself until the middle of August. It would come in a little bit in the morning, and now we begin to see that in the middle of July. Now we begin to get showers in late August and September, and that has really impacted the harvest season since we grow so much seed and grain. It's become difficult to get those crops dried down so that we can harvest them." Much of the "produce" on Nash's farm comes in the form of seeds—livestock feed, food grains, cover crop and vegetable seeds grown on contract for commercial seed companies. "Our seed crops have become very difficult because of the instability in September. We lost a beet seed crop last year, a very valuable seed crop, because of the rain in September."

More variable weather has not caused a change in the crop mix on Nash's farm, but has forced him to buy more tractors, tools, processing equipment and combines to take advantage of increasingly narrow windows of time when conditions are right for planting, cultivating for weeds, applying compost or harvesting crops. "We switch out crops depending on how the weather is affecting our ability to do fieldwork. For example, if we can't get into the field early enough in the fall, we put in barley instead of wheat. It's very, very quick decision-making. We have a generalized pattern, and then we switch up the crops depending upon what fields we can get into, when we can get into them and how much of each grain that we need."

Like most growers in the Fruitful Rim, Nash irrigates all of his crops. "We have a very sophisticated irrigation system here on the north peninsula. The mountains store the snow, and then in the spring and the summertime, we use that water for irrigation. Last year [2013] my farm got five and a half inches of rain the whole year. Our normal is around twice that. We get good spring snow in the mountains, which allows us to irrigate crops all summer long. If we don't get that snow, then we're in the same problem that they have in California now."

Thinking about the future, Nash wonders how warmer winters and competition for water might affect his farm and others in his region. "If we don't get the snowpack, we don't have water. Change in water access is one thing that could really impact the farm—it is the 900-pound gorilla in the room that nobody is talking about."

Over the years, Nash's Organic Produce has won national recognition for conserving natural resources while growing abundant high-quality food, including a Steward of Sustainable Agriculture or "Sustie" award from the Ecological Farming Association in 2011 and a Steward of the Land Award from the American Farmland Trust in 2008.

The Southwest

Paul Muller, Full Belly Farm, Guinda, California

Full Belly Farm, in the Capay Valley of Northern California, is a 400-acre diversified organic farm raising more than eighty different crops, including vegetables, herbs, nuts, flowers, fruits, grains and livestock. The farm landscape is home to a diverse interweaving of perennial orchards, annual crops and pastures, plus hedgerows and riparian areas managed as habitat for beneficial insects, native pollinators and wildlife. First established by Paul Muller and Dru Rivers in 1984, Full Belly Farm has involved an active partnership since 1989 among four owners who live in three households on or close to the farm: Paul Muller, his wife Dru Rivers, Judith Redmond and Andrew Brait.

Full Belly Farm was designed to be ecologically diverse to foster sustainability on all levels, from healthy soil to content consumers; a stable, fairly compensated workforce; year-round cash flow and an engaging workplace that renews and inspires everyone working on the farm. The productivity of the diversified organic system is based on the use of cover crops and the integration of sheep and poultry to capture and cycle crop nutrients and water, maintain soil health and prevent losses

FIGURE 5.2. Judith Redmond, Andrew Brait, Dru Rivers and Paul Muller, Full Belly Farm, Guinda, California. Credit: Paolo Vescia

from pests and disease. Virtually all of the production on the farm is irrigated, mostly with water from Cache Creek, which runs along one side of the property.

The farm sells to a diverse mix of direct and wholesale markets in the San Francisco Bay area that includes restaurants, grocers, farmers' markets and a 1,500-member CSA. Full Belly also supports a number of outreach programs to help create awareness of the importance of farms to all communities.

According to Paul Muller, three years of extreme drought coupled with dryer and warmer winters, longer and more variable spring and fall seasons and greater weather extremes have created both opportunity and challenge on Full Belly Farm. "The last couple of years we have had the driest January and February on record. That's the time of year we normally get the moisture that goes deep in the ground, the moisture that serves as the reservoir for our crops that come in the spring. They get their roots down deep and draw on that water. The last couple of years, it just hasn't been there."

Declining water supplies have made growing the cover crops so crucial to building soil quality and providing nutrients for crops more challenging. Increased weather variability has made planning and conducting fieldwork more difficult by reducing the periods when work can be done without damaging soils and crops. Paul explains, "Normally we can't prepare ground for planting in February, but it is so dry now that we can. The challenge is that you've got to get your groundwork done as early as possible and when the time is right. We used to get light rains in fall that moistened things up and made large windows to prepare ground, but now variability has narrowed windows and made them less predictable."

The longer, more variable springs and falls have complicated crop management, but also created opportunities by increasing the length of time the farm can produce valuable spring and fall crops. "The way I look at it is, we are gambling more on the edges. Summer temperatures have been relatively stable, so we can hit our main summer season, but now our main season shifts a couple of weeks forward or backward, but those summer crops—melons, tomatoes, beans, peppers—remain the same. If spring weather is unseasonably hot, our spring will end sooner than we thought. If our summer is unseasonably long, then tomatoes will go longer, but the fall crops won't do quite so well because we did not get them planted early enough. And maybe in mid-winter, the January-February crop mix, if it's unseasonably warm, we'll harvest more crops in those months than we normally do. Our crop mix is not changing a lot, but we're more conscious of how we plan for the edges." Paul goes on to say that increased weather variability means the edges can have "bigger bounces" and more extreme swings, but that the farmers can push the edges, and sometimes they do really well.

> A lot of our practices are focused on conserving water. How do we capture the water that does fall on the ground, so it doesn't run off, so that it goes into the soil? How do we make the water that we have go further? Water is the critical piece here in California.
>
> — Paul Muller

Although water conservation has always been a management focus at Full Belly Farm, heavier rainfall, longer dry periods and continuing drought have encouraged even more thinking about sustainable water management. Paul says that the management team is considering changing up their crop mix to include more drought-tolerant cover crops and is exploring potential cover crops that do better on less water or produce more with the same amount of water than their current cover crops. They are also looking at ways to use cover crop mulches to conserve soil moisture and are weighing the costs and benefits of more water-efficient irrigation systems such as drip and microsprinklers; these involve significant initial investment and add management challenges because they require filtered water, which the farm does not need now.

The managers are also working on landscape-scale improvements in water management. The main water source, Cache Creek, drains a large watershed above the farm. The low fields near the creek are kept in winter cover crops to reduce soil loss in the event of a flood, which would be most likely to occur during heavy winter rains. The farm also actively manages riparian zones along the creek so that when it overflows its banks, flood waters will move over the lower parts of the farm without damaging production areas.

Looking to the future, Paul expressed a number of concerns about the lack of coordinated planning for agricultural adaptation to climate change in his region and elsewhere. He also believes that the recent spike in farmland prices in Northern California is related to climate change. "Investors and growers are moving into more water-secure areas if they can. Other growers are trying to secure water access in other ways. For example, many are investing in deeper wells, even though that reduces water available to other farmers. There is a collision of interests that farmers are starting to pay attention to."

5.1. Can California Avoid a Tragedy of the Commons?

Human nature compels the unsustainable use of natural resources held in common even when the costs of continued exploitation are clear. California agriculture is at the center of a collision of growing demands for an increasingly limited resource—clearly a "tragedy of the commons" in the making.

Sustainable management of the state's water resources has been hindered by a growing population, new concerns about environmental degradation, an arcane web of local and regional water regulations, aging flood control and water distribution infrastructure and the uncertainties of a changing climate. Without comprehensive reform, California is likely to experience widespread economic and environmental damage.

The California Water Action Plan, released in 2014, is designed to provide more reliable water supplies, restore important species and habitat and establish a more resilient and sustainably managed water resource system for farms, ecosystems and communities through funding for improved water efficiency, wetland and watershed restoration, conservation and integrated water management.[1]

Paul goes on to wonder if all this planning and investment by the farming community will be for nought if other water users—urban and industrial—take control of water supplies. He suggests that Northern

California could reap huge dividends from an investment in a coherent rangeland management strategy designed to improve the health of the regional water cycle.

According to Paul, the new FDA food safety rule[2] conflicts with current water recycling programs and could further restrict water availability in his region. "A lot of the FDA food safety rules are predicated on the assumption of an abundant resource, a stable water supply and a stable climate." Paul would like to see more programs with NRCS to look at climate threats and develop strategies to create a balance of ecological health and farm protection. "There was a study from Stanford a number of years ago that said that agriculture is being overlooked as a sector that, one, has the largest impact on land use in the country, but two, doesn't have a coherent strategy as to how they address rising CO_2 levels. So maybe under that corn in the Midwest, there could be clover growing, so that as that corn's done, the clover's there. And it's sequestering carbon all fall, and into the spring it starts again. And maybe we invest in other equipment that will allow that to happen, rather than just using herbicides. Let's take a serious look at polycultural systems and other multilayered systems that provide greater ecological services in terms of sequestering carbon and enhancing climate resilience, and let's pay farmers for using those practices."

The farm's co-owners have collected numerous awards over the years for their success in creating an exemplary model of sustainable agriculture. In 1999, Paul and Dru were named Outstanding Farmers of the Year by the University of California Small Farm Program. Full Belly Farm also won a Steward of Sustainable Agriculture or "Sustie" award from the Ecological Farming Association in 2005, the Patrick Madden Award from the USDA in 2006 and a Growing Green Award from the NRCD in 2009.

Brendon Rockey, Rockey Farms, Center, Colorado

Brendon and Sheldon Rockey are the third generation to grow potatoes on 500 acres of irrigated land in the San Luis Valley of Colorado. The Rockeys operate a farm and packinghouse near Center. Brendon is in charge of field operations while Sheldon is operations manager of the business which packs specialty potatoes for certified seed and fresh table use. The farm focuses on direct sales to commercial potato growers and the wholesale fresh table potato market. They also do some direct sales from the farm.

Brendon has successfully incorporated sustainable agriculture practices like cover cropping and companion planting into the production system to improve soil quality and conserve water. The increased soil quality that Brendon achieved with the new cropping practices improved farm profitability because he found that he could use less water and decrease or eliminate fertilizer and pesticide use while maintaining yields and improving crop quality.

Brendon Rockey says that concerns about water use on their farm pushed him and his brother into trying cover crops. Average rainfall in the valley is about five inches per year, so all agriculture there is irrigated. Growers pump from wells, and that water is replaced each spring from the snowmelt, which flows down from the mountains to replenish the groundwater. During this drought, growers have been pumping more water from the aquifer than is being replaced by the snowmelt, and in 2013 the aquifer level hit a record low. "Water has always been a huge issue for us out here in southern Colorado. We're in the middle of a drought like a lot of America, and it's been going on now for fifteen plus years. It has really forced us to make some changes to our management practices, but it's actually helped out our farm as a whole," explains Brendon.

FIGURE 5.3. Brendon Rockey, Rockey Farms, Center, Colorado. Credit: Brendon Rockey

Longtime growers of certified seed potatoes as well as fresh market potatoes, the brothers decided to drop barley from their two-year potato/barley rotation about ten years ago and replace it with a mixed cover crop as a way to decrease water use. They reduced their water use by about fourteen inches, and the switch had some other significant and unexpected benefits. As

Brendon explains, "Bringing in a diverse cover crop improved our soil health so much that it had a huge impact on the productivity of our potato crop." The increase in soil quality reduced input costs and increased potato quality so dramatically that Brendon found it was more profitable to grow one cash crop every two years than one cash crop every year.

The cover crop success and the extreme drought in the Southwest has Brendon thinking about how to get even more out of his cover crops. As the drought continues unabated, Brendon is contemplating a polyculture designed to encourage beneficial insects and suppress pests in the potato crop. "We've seen so much positive impact from having the multispecies out in the cover crop that we are thinking about bringing more diversity into the potato crop. Next year I am planting a three-species companion crop and an eight-species insectary crop in the potatoes. I'm planting peas, chickling vetch and buckwheat in the rows with the potatoes. So I've got the two legumes out there for my nutrient management and the buckwheat attracting insects. I am also going to plant an occasional row of insectary mix in among the potatoes as well for the purpose of attracting predatory insects." Brendon hopes to further reduce input costs and increase soil quality with the additional diversity added to his crop rotation.

For every research paper you read on global warming, you find another one saying it is getting cooler. I think weather cycles, but I don't get too hung up on patterns because it is beyond my control. My whole focus is just creating a resilient system that can handle climate change, whichever direction it might be.

— Brendon Rockey

Looking ahead, Brendon wonders about the future of agriculture in his valley. If the water supply becomes even more limited, growers will have no choice but to start taking acreage out of production. "I guess that's the real scary thing. I'm hoping that we can get enough guys to do the right thing and save enough water that we don't get to that point. It seems from the outside like it would be easy, but the attitude here is like, 'I wish the neighbors would all cut back on the water so I can keep farming every acre I have.' You just try and get a bunch of farmers together and get them to all agree on the same approach. It's really difficult!"

Brendon and Sheldon regularly host visitors and lead workshops at their farm, and Brendon is a regular speaker at farming conferences and workshops in the Southwest. The brothers were nationally recognized in 2012 for their innovative potato production system as recipients of the Soil and Water Conservation Society Merit Award for promoting sustainable agriculture and soil health.

Jacquie Monroe, Monroe Family Farms, Kersey, Colorado

Jerry and Jacquie Monroe are the third generation to farm his family's 20-acre "homeplace" in Kersey, Colorado, about an hour northeast of Denver. Monroe Family Farms is the oldest organic farm in Colorado. When Jerry and Jacquie took over from Jerry's father in 1991, they went into the business of growing organic vegetables in a big way, adding 175 acres and starting the first CSA in Colorado in 1993, because they wanted to work closely with people who appreciated their farming philosophy. Today, the farm produces a hundred different kinds of vegetables and all the pasture, hay and feed grains needed to produce pasture-based meats (beef, pork, and lamb) and eggs onsite—all of it USDA-certified organic. With the help of seven employees, Jerry manages the crop and livestock production while Jacquie manages sales and distribution for their year-round 650-member CSA. The farm also markets to select restaurants in Denver and Boulder.

FIGURE 5.4. Jacquie and Jerry Monroe, Monroe Family Farms, Kersey, Colorado. Credit: Jacquie Monroe.

The Monroes emphasize soil health, water and energy conservation on their farm. They maintain soil health by integrating livestock into a diverse rotation of vegetables, alfalfa and feed grains. Irrigation has been upgraded from gravity-fed, furrow irrigation to more water-efficient pivot and drip systems, and they use tailwater ponds to capture and return to the fields any surface runoff that occurs during irrigation or rainfall events. Produce for the winter months of the CSA are stored in dugouts, pits and straw-bale buildings to reduce energy use. Jerry and

Jacquie have succeeded with a philosophy of growing ample quantities of organic, life-filled and healthy foods while conserving and respecting the natural environment and providing an educational experience working with Mother Nature for any CSA members who want it.

Sixteen years of extreme drought combined with higher summer temperatures, warmer winters, more extreme weather and a longer growing season have put the focus on water efficiency. "Water management has become huge," says Jacquie. "Jerry has to keep track of how much water we have, how much he's used. Everything has been flood irrigated here. Back in the day, they dug ditches and then had these pipes that went over the ditches, and fields were just flooded to irrigate them. When you flood irrigate, the top of the farm gets more water and the bottom of the farm gets less water and in the middle is the only part that gets the perfect amount of water."

We here in Colorado have been in and out of a drought since 1998 and more in a drought than out so water out here is everything. We have to irrigate in order to get a crop, so water is a huge problem especially since most of the fresh water is owned by the farmers and all the cities are taking that water away.

— Jacquie Monroe

"Then we had to start conserving water because of the drought, so we started putting in drip irrigation. I would say we have sixty acres of vegetables. When we first started with the drip irrigation, we put ten acres in. We're up to probably forty acres of drip irrigation. And the rest of the crops are now under a new center pivot that we put in just this year because of the shortage of water. We want to make sure that we can continue to grow vegetables, and the pivot and drip irrigation puts that exact amount of water throughout the whole entire system. We are doing a lot more with water management. I don't know what we would have done if our weather hadn't changed and we hadn't been so dry."

Jacquie says that another change that has come with the drought has been more challenging weeds. "Weeds are starting to go crazy out here. We're finding some of them are becoming very invasive. I can give you two examples. We've always had what's called goatheads. It's a small weed that grows very low to the ground and has a burr that sticks in your tires and in your shoes. It used to be only in certain parts of the farm, but now it seems to be going everywhere. The other one is sunflowers. We've never had sunflowers here before the drought. They are literally taking over all of our ditches. Anywhere that you can't mow or get to, they grow like trees, and we can't seem to get rid of them. These weeds are getting to be a problem for us. We're trying to do a lot of mowing to try to keep them down, but the darn things adapt. They're growing shorter now and growing a head and flowering close to the ground where the mower doesn't hit."

Like many other vegetable growers across the nation, the Monroes have also found opportunity in the longer growing season created by the changing climate. Jacquie estimates that in the last decade they have extended their growing season nearly two months with the help of some physical protection for frosts. They are in the field about a month earlier in the spring and can extend the fall harvest season about a month longer than they used to. "We're picking things by the first of June, and that's never happened in our lives. That is crazy to think that we are able to produce something and harvest it by the first of June when our official last freeze date is May 15. We've extended our harvest season, and our income has increased because of it."

Jacquie and Jerry's twenty-two-year-old son is weighing the pros and cons of joining his parents in the farm business. He would like to become the fourth generation on the farm, but competition for water in the region makes it difficult to imagine a lifetime in farming. "We are very concerned in the future about our water rights and whether or not we're going to be able to get our water," says Jacquie. "The cities are buying the water off the farms and taking it back to the city so people can eat, drink, bathe and water their lawns. Our elderly farmers are selling out. Once a person has gotten to a certain age and there isn't a family member who wants to take over the farm, they sell their water. I don't blame them. They finally have something of value that somebody wants and they're paying them well for it, but it sure hurts the rest of us."

"I'm to the point where I'm going to start asking at our annual water meeting that the farmers quit selling it, that they rent it to the cities for the rest of their lives and the rest of their children's lives so that we can keep control of our water. That way the farmers can still have some kind of control over what's going on out here. The cities are drying up our farms. They say that seven hundred thousand acres is supposed to be dried up in the next ten or fifteen years. It means that water will never go back to those farms. Once it's gone, it's gone forever."

The Midwest

Richard de Wilde, Harmony Valley Farm, Viroqua, Wisconsin

Harmony Valley Farm is a diversified farm that spreads out over 200 acres of cropland, pastures and forest near Viroqua in southern Wisconsin, about two hours northwest of Madison. Richard de Wilde and Andrea Yoder, the co-owners manage the production of about 100 acres of organic vegetables and berries at the farm and on some leased land nearby.

Richard is a cofounder and master grower at Harmony Valley Farm, established in 1985, and applies his forty-plus years of farming experience to the integrated management of a healthy natural growing environment on the farm. Richard has always made managing for soil health a priority, believing it to be a key contributor to the success of the farm. Over the years, he has developed a system of cover cropping with green manures, applying natural rock powders and incorporating compost to maintain healthy soils. He controls pests by managing perennial habitat and nesting sites for beneficial wildlife including raptors, songbirds, bats, wasps and insects. Harmony Valley Farm is best known for its season-long, high-quality salad mix, sauté greens and spinach, as well as root crops harvested in the fall and distributed throughout the winter. The farm also produces grass-finished beef using intensive grazing practices.

FIGURE 5.5. Richard de Wilde, Harmony Valley Farm, Viroqua, Wisconsin. Credit: Center for Culture, History, and Environment, NIES, U. of Wisc.-Madison

Harmony Valley Farm sells organic produce, berries and beef through direct and wholesale markets, including a 1,500-member CSA that runs from May through January with deliveries locally and to Madison, WI, and Minneapolis and St. Paul, Minnesota. The farm also sells at the weekly Dane County Farmers Market, and to retail grocers and wholesale distributors throughout a large area from the Twin Cities to Madison. It employs a large team that varies from fifteen to sixty members, depending on the time of year, to produce and market its products.

Starting about seven or eight years ago, changing weather began to require some changes in production practices at the farm. More frequent heavy rains and stronger winds, more variable springs, warmer summers and longer falls have complicated vegetable production, according to Richard. "River bottom land is the best kind of land for growing vegetables in our area. And it's great in dry periods, because we can irrigate out of the river."

But the farm's million dollars in losses in 2007 and 2008 as a result of unprecedented flooding really got Richard's attention. "Not many people understand that vegetable farmers have little to no insurance against weather.[3] We can participate in the USDA's NAP program and we do. N-A-P is the abbreviation for Noninsured Agricultural Production. Noninsured meaning it's not corn, soybeans, cotton, wheat. It's not a commodity. NAP is a poor program. It is. It's totally inadequate, and we really don't have much else. After the flood in 2007, USDA did not help us out. But if you're a corn farmer, you can buy government-supported, 90 percent-guaranteed income on the corn crop. It's gross. We should care more about feeding people than raising corn for export and ethanol and corn syrup."

In August of 2007, we got hit really hard with some really weird flooding caused by 18 inches of rain in a less than a 24-hour period. A lot of crops were peaking just then, like tomatoes. We had pretty big losses because a lot of our farmland is along the Bad Axe River. They called that a thousand-year event. And then we had another one nine months later. That was when I said, "There's no such thing as normal anymore."

— Richard de Wilde

Richard remembers when weather used to move pretty predictably from west to east. During the 2007 flooding, he noticed for the first time a weather pattern that he associates with severe weather. "Something that I had never seen before the 2007 flood is a pattern of southerly flow bringing moisture up the Mississippi River Valley. The moisture turned in a circle before it hit the Great Lakes, and then it just looped back on us and didn't stop. It just didn't move off, and that's why we got eighteen inches of rain. Now we have these weird looping events. I've seen it several times since, and now it just scares me when I see that loop."

Other than the extreme flooding events, most of the changes in weather Richard has observed are more severe expressions of familiar seasonal patterns. For example, extreme swings in the timing of spring are more common these days. For many years, spring planting at Harmony Valley Farm began reliably in the first week in April, but in the last decade, it has begun to vary by almost a month, which increasingly complicates spring planning and transplant production. Falls generally are longer but more variable, so Richard has extended the fall production season "knowing that we are going to get burned sometimes."

Richard also has noticed that winds seem to have become more frequent and intense. "Strong winds have definitely become more of a factor in the last few years. We have more wind and stronger wind. We lose row covers more often now. We've had more problems with row covers—even if they'd stayed on, there was so much wind that the movement of the cover abraded the leaves, and so we have crop damage even under the cover, even if the cover stays on. In the winter, it used to be that the winds died down at night. This winter we've had an amazing amount of night winds, and that brings more risk of wind chill."

Richard de Wilde has received national recognition for his long record of success as an organic grower and as a CSA marketer with a "Sustie" award from the Ecological Farming Association and a Farmer of the Year award from the Midwest Organic and Sustainable Education Service, both in 2003.

The Northeast

Jim Crawford, New Morning Farm, Hustontown, Pennsylvania

For more than forty years, Jim and Moie Crawford have owned and operated New Morning Farm, a 95-acre certified organic vegetable farm in south-central Pennsylvania. Jim and Moie manage a diverse mix of vegetables and small fruits—about fifty different kinds—on 45 acres of the farm's best cropland. They also sell eggs produced on the farm.

New Morning Farm has a greenhouse and four high-tunnel cold frames for transplant production and season extension, access to ample surface water for crop irrigation with sprinkler and drip systems. Soil quality is maintained with a diverse crop rotation that puts about a third of the land in cover crops each year and includes the regular application of purchased and locally made composts and plastic mulch. The farm employs six to eight year-round employees and twenty-five seasonal workers, including those participating in the farm's well-respected apprenticeship program.

FIGURE 5.6. Jim Crawford, New Morning Farm, Hustontown, Pennsylvania. Credit: Jim Crawford

About two-thirds of the farm's produce is marketed directly to consumers at markets in Washington, DC, while one-third is sold wholesale through Tuscarora Organic Growers, a marketing cooperative that Jim helped to organize in 1988 to coordinate the sale of fresh fruits and vegetables to retailers, restaurants and institutions in Washington, DC.

Growing vegetables on bottom land in central Pennsylvania has brought Jim Crawford a lifetime of intriguing challenges, most of them related in one way or another to water. "I can remember my very first year farming, the big issue was it just kept raining all spring. The dry days were so few that we just couldn't farm. People said, 'Oh that's just Pennsylvania. That's just the way it is here.' Basically, our biggest challenge was to figure out how to farm when it was always raining." Paradoxically, learning how to manage crop irrigation during frequent summer dry spells typical of the region was another big challenge. According to Jim, "irrigation in this part of the country is a way bigger factor than many people realize. It's just a huge issue for yields, quality and profitability, and production in general. Irrigation is an enormous challenge."

About fifteen years ago, Jim began to notice shifts in weather on the farm. Temperatures seemed to be getting more variable, and there was a definite increase in both heavy rainfall and summer drought. These changes seemed to intensify after 2010. "It's really hard to draw conclusions, but the variability just seems so extreme in the last few years. We saw 80 degrees in April, in March and then in January. Those highs are then followed by real, real cold temperatures. Just extreme shifts in temperature, extremes you did not used to see so often. And there has been more drought and more flooding. We're going through a drought right now that's one of the most extreme that I've ever seen in forty years. I'm not complaining because we have been able to irrigate. It's just that I think it's another sign of extremes."

> I realize now how lucky we were for so many years. We didn't see many of the classic vegetable diseases at all for most of our years. But in the last 5 to 8 years, things have really changed. Why is it that we suddenly have seen a whole range of really devastating diseases that we never saw before?
>
> — Jim Crawford

One key to Jim's success in managing vegetables in the typically wet conditions on his farm is the use of black plastic mulch as physical protection. The plastic mulch allows Jim to do much of his fieldwork during the fall and winter, when conditions are more often right for incorporating soil amendments and shaping planting beds. "What we have done now for six or eight years is to prepare beds and put plastic down on large acreage before we need it. That way the land is standing by, and it's ready when we need it. So the tillage is done, the spreading is done. Of course, we get weed control and moisture control out of it too, but the biggest thing it does for us is it keeps us on schedule."

Like many other farmers, Jim has used the lengthening growing season to his advantage. For example, with the addition of physical protection in the spring and fall, he has been able to extend his harvests of sweet corn and snap beans. "We have actually doubled the number of weeks that we harvest sweet corn and more than tripled the weeks of green beans. We've drastically increased the length of our season for those two crops." But Jim acknowledges that improved season-extension practices have played a large role in lengthening harvests too. "It's definitely not all weather. We've also had big successes on season extension with various techniques. But it does indicate that the growing season is longer. And I have to say we have made money on it, even though it makes me very uncomfortable because I know it is a sign of climate change."

Another sign of a changing climate may be the increasing costs and management challenges created by a growing number of novel plant diseases. Late blight arrived on the farm for the first time in 2003 and has been a frequent disease problem in Jim's tomato crop since then, significantly reducing yields in four of the last ten years. Another recent arrival to the farm is downy mildew. "We didn't see downy mildew in cucumbers until 2007. Now we see it every year, around the first of August. It wipes out the cucumbers. No more cucumbers. It's really dramatic. It's amazing to me. We used to have a full cucumber season from June to October, and I can remember beautiful crops of cucumbers in October. Cucumbers lying everywhere. Now, we don't see a cucumber here after the beginning of August."

Other costly new diseases that have arrived on the farm in the last decade are *Alternaria* in brassicas, a mildew in basil and a rust in their raspberries. Jim explains, "We grow a lot of basil, it's one of our bigger crops. It's in our top ten of the fifty crops that we grow. We went for decades never having a disease problem with the basil, and now we have disease all the time, every season, all season long. But the biggest new disease is this rust in our raspberries, which are a pretty big crop for us. We never saw it until last year when it wiped out the whole crop starting in July. We thought we were on top of it this year, and now here just this week, it's back and it's wiping us out again. Thirty-eight years, we never saw it. Now we've seen it two years in a row. It isn't a little problem. It destroys the crop."

Central Pennsylvania is in the path of Atlantic hurricanes, which make a regular appearance on the farm. "Hurricanes are the single weather events that have definitely cost us, by far, the most money over the years," says Jim. "They usually come in September. That's been our most vulnerable times. Although we've had them through the years, the biggest ones ever have both hit in the last decade. Both came in September, and both wiped out probably a third of our year's production all in one day." Jim has investigated insuring his crops through the USDA, but found that the programs available to him were not appropriate to his farm because of his crop diversity and the value of his crops relative to conventional vegetable farms.

Jim regularly leads farming workshops, gives lectures and hosts field days at the farm for farm organizations and local colleges and universities. In 2002, the Crawfords were recognized for their long and active support for organic farming with a Leadership Award from the Pennsylvania Association for Sustainable Agriculture.

Elizabeth Henderson, Peacework Farm, Newark, New York

Elizabeth Henderson has grown organic vegetables for more than twenty-five years in Wayne County, New York, and is a founding member of Peacework Organic CSA. Peacework is located on 109 acres of prime farmland that has been protected from development in a unique community collaboration that enabled the Genesee Land Trust to purchase the farm and then lease it back to the farmers on a twenty-five-year rolling lease. The farm grows more than seventy different crops on 20 acres, including a wide variety of vegetables, herbs and flowers.

FIGURE 5.7. Elizabeth Henderson, Peacework Farm, Newark, New York.
Credit: Elizabeth Henderson

The produce is marketed through a 300-member CSA over a six-month harvest season.

Designed to maintain soil quality and reduce pest pressures, cropping systems on the farm focus on summer and winter cover crops, organic mulches and some purchased compost. A greenhouse and hoophouses are used to produce vegetable transplants and extend the season in the spring and fall, and ample water is available for drip irrigation from three wells on the farm.

Managing the effects of heat, drought and excessive rain on the crops and people of Peacework has got Elizabeth Henderson wondering about the challenges facing new farmers these days. Hotter summers, heavy rainfall, drought and novel diseases have required some adjustments to the farm's management practices. "When we started farming, we didn't have any irrigation. We could rely on the rain and the soil's organic matter to get us through. There would sometimes be a couple of dryish weeks, but then there would be rainstorms in July, you know, thunderstorms and we could get through a season. Since about 2000, no two years have been alike. Really wet years, dry, dry drought years—and you never know at the beginning of the season what to expect. With the really erratic weather, we found that we had to install trickle irrigation and dig a well in each field so we could have reliable water. And then when you have to use irrigation, about a quarter of your time goes in to managing that."

Elizabeth experienced her first severe drought as a farmer in 2005, a year that the *Farmer's Almanac* predicted would be average rainfall. "There was a quarter inch of rain between the beginning of May and the third week of September that year. One quarter inch! We had never had that before. And then it rained every day in October, so it WAS an average rainfall year!"

Higher summer temperatures and particularly heat waves have challenged both the plants and people at Peacework. "People who live in the South are used to really hot weather, but my partner and I are Northerners and heat just makes us totally miserable. We just have never experienced weather of over two weeks in a row over 90 until the past two years. That just makes the working conditions extremely difficult, and of course you get up earlier so that you can try and do harvesting during cool weather, but when it never goes down below 90 degrees, there isn't any cool weather for really good harvesting. It's much harder on our people. It's grueling."

Elizabeth is thinking about how to apply some of the crop management practices she observed in South Korea and Taiwan to manage crops in the heavy downpours that have become more frequent at Peacework. "It's getting harder to manage water on our farm. When rain comes, it isn't just gentle rain, it comes in two- or three-hour downpours of three inches. When I visited Taiwan the first time, I noticed that they have tomato trellising that was three times as sturdy as I thought you would need, not just a tripod with three bamboo poles, it would be a tripod with eight or nine bamboo poles. And then I saw a typhoon, and I saw that their tomatoes were still standing, because they have learned how to make trellises that can stand up to a typhoon."

"The kind of rain I saw while I was in South Korea and Taiwan explained why they use hoophouses in the summertime to protect their crops. Rain coming down so hard that it washes the crop out of the ground, or flattens it. Or if it's seed, it could bury it or wash it out. They've been managing heavy downpours a long time." Peacework has four hoophouses that have been used in the past for season extension and disease management, but recently there is talk about how to use them to protect crops from heavy rains.

> When I started farming, there was a pattern to the weather, one year was kind of like the one before. I think it's more challenging to be a farmer now. No two years are alike. You have to be more flexible, you can't rely on a plan. You have to learn as many tricks as possible, because it might be dry or it might be wet. You have to be so nimble these days.
>
> — Elizabeth Henderson

There is also discussion about returning to some water and wind management practices that were used in the past. "We are lucky that we have sandy loam over gravel. We used to have it set up with grass strips between the beds, and that was a pain in the neck because you had to mow those strips. So we took most of them out, and I'm thinking that was a mistake. With the grass strip, you can get on a bed way earlier, because the whole field doesn't have to dry out. You can ride your tractor on the grass strip and do light tilling of a bed, where it would be too wet if there weren't the grass strip there." Because the Peacework property is level, water erosion is not a concern, but the soils, plants and people are exposed to wind. "In a very dry summer, there can be a lot of real painful wind erosion, with the wind whipping through the sand. So I think being careful to have grass strips and windbreaks of bushes or trees is essential for a changing climate."

Like other growers in the humid East, Elizabeth has seen a startling increase in crop disease with the changes in weather over the last decade. "In my first fifteen years of farming, we never lost an entire crop to a disease. You would have some disease on some of the crop, or some pest, but in the past ten to fifteen years, we've had things like powdery mildew blow in and entirely wipe out all the cucumbers. Or late blight totally wipe out the tomatoes and potato crop. That was just not an experience that I'd had before."

Elizabeth expressed concerns about the lack of federal crop insurance programs for diversified vegetable growers faced with managing the novel risks she has experienced from climate change effects on her farm. "The crop insurance that is available is not appropriate for our farm. It's getting better. They have some new policies that are more accommodating to a farm growing seventy crops, but for the most part, up until now they haven't had that. You get the insurance, you pay by the crop. It hasn't been a good fit. And it is very irritating to me, that despite the hazards for farmers, insurance companies consider crop insurance really a nice profit area, because they get subsidies from the government. They are subsidized by the federal government. Their guaranteed profit is higher than any farm ever gets."

Looking to the future, Elizabeth believes we have the knowledge we need to sustainably manage soils and crops in a changing climate. She believes that efforts to improve the adaptive capacity of farms are best focused on the human dimensions of agriculture. "I think we've paid plenty of attention to soil and crops and not enough attention to people. The past ten years I've been working on the Agricultural Justice Project,[4] which addresses issues of domestic fair trade. It's developing standards for fair payments to farmers that fully cover production costs and fair labor standards on the farms, so the people who actually do the farm work are treated better, given more respect and are involved more in the farming. I think that's what we need to understand better because in these very, very challenging times, having one manager is not enough. Everybody on a farm has to be constantly observing, trying to understand and working out how to be nimble together."

Elizabeth has been an active organizer, advocate and author for organic farming, sustainable agriculture and food justice for many years and represents NOFA on the Board of the Agricultural Justice Project, which offers Food Justice Certification. She has been nationally recognized for her work as a recipient of the Spirit of Organic award (2001), Cooperating for Communities award (2007), a NOFA-NY Lifetime Achievement Award (2009) and, most recently, the Ecological Farming Association's Justie Award (2014). She co-authored the book *Sharing the Harvest: A Citizen's Guide to Community Supported Agriculture* (Chelsea Green, 2007), which has helped many farms connect directly with supportive customers willing to share the risk with their farmers. Although Elizabeth retired from full-time farming in 2012, she remains active as a mentor, food justice activist and author.

The Southeast

Ken Dawson, Maple Spring Gardens, Cedar Grove, North Carolina

Ken Dawson has raised organically grown vegetables for more than forty years in the community of Cedar Grove, North Carolina, located about twenty-five miles north of Chapel Hill. Ken and his wife, Libby Outlaw, established Maple Spring Gardens in 1983 on leased land and moved their farm business to a worn-out tobacco farm in Cedar Grove that they purchased in 1990. Their long experience growing for high-value markets and using sustainable practices like composting and cover crops to build and maintain soil quality swiftly transformed the badly neglected land into a productive and profitable farm.

Today, Ken uses crop diversity, crop rotation and cover crops to maintain soil quality and reduce pest pressures and insectary plantings and OMRI-approved pesticides, when needed, to manage pests and diseases. Maple Spring Garden has a 5 kW photovoltaic array tied to the grid and ample water for irrigation from a pond and a well on the farm. With the help of a seasonal crew of six to eight fieldworkers, Ken and Libby grow more than eighty different varieties of vegetables, cut flowers, small fruits and medicinal and culinary herb starts on 6 acres of seasonal production rotated through 14 acres of cropland. They market their produce through direct sales to a 200-plus member CSA and at farmers' markets in nearby Durham and Carrboro, and to local businesses.

The Southeast has always been a difficult place to grow vegetables. Poor-quality soils, fluctuating winter temperatures and hot, humid summers encourage pests and diseases and reduce crop yields."It's my perception, and I certainly don't have the weather records to document it, but my perception is that the variability is becoming greater both in temperature and precipitation. Obviously the variability in precipitation is always a challenge, but the more extreme it gets, the more of a challenge it is. For, example, in 2002, we had the driest year in a hundred years in central North Carolina. In 2003, we had the wettest year in a hundred years in central North Carolina. In 2007, we had the driest year ever recorded in central North Carolina. This year, we had the coolest, wettest season anybody living can remember."

> The frequency and intensity of extreme weather seems to be increasing, which presents all sorts of challenges, the unpredictability of it. The last year that I remember as what I would consider a really good growing season was 2001. Since then, we've seen it all. We've had the driest years ever and the wettest years ever and the coldest winter in decades and the hottest summer in 100 years. The extremes are just becoming more extreme.
>
> — Ken Dawson

"Likewise in temperatures. The winter of 2010 was the coldest year in thirty years here, and that summer was the hottest summer ever, with July and August just constantly setting temperature records for most days...hitting 105 for days on end here and the most 90-degree days ever recorded. Variability is always a challenge, that is a given in farming. It's not like we're seeing things we haven't seen before, just more so. The high temperatures, the heat waves, it's all just seems to be becoming more extreme. The extremes are just becoming more extreme."

Hotter temperatures and more frequent heat waves have begun to interfere with crop production on Ken's farm and others in the region. "Flowering, pollination and fruit set had never been an issue for us prior to 2010. In the 2010, '11 and '12 growing seasons, we had very poor fruit set on our late tomatoes due to excessive heat in July. That's something I had never encountered before. Early September, when we normally have a lot of late tomatoes, there just weren't any. They were great-looking plants with nothing on them." The late crop of tomatoes is an important crop for many growers in the region, including Ken, but many have now given up on the crop. Ken is thinking about trying some heat-tolerant cultivars for the late planting.

FIGURE 5.8. Ken Dawson and Libby Outlaw, Maple Spring Gardens, Cedar Grove, North Carolina. Credit: Debbie Roos

5.2. High Temperatures Disrupt Vegetable Production

Lettuce is a cool season crop that is particularly sensitive to variable temperatures and moisture. For optimum growth and crop quality, daytime temperatures should remain between 55 and 85 degrees Fahrenheit and nighttime temperatures between 40 and 60 degrees.

If temperatures rise above this range, even for a short time, lettuce will flower prematurely. This response is called "bolting," and in lettuce it is accompanied by a lengthening of the stalk and the production of bitter compounds in the leaves, which greatly reduce quality and marketability. A sudden drop to freezing temperatures can kill young plants and cause blistering and peeling of exposed leaf surfaces on mature plants, which can cause decay during handling and storage.

Most warm-season vegetables are plant fruits—they contain the seeds of the next generation—so successful production requires environmental conditions that support healthy flowering, pollination and fertilization. Extreme temperatures during the reproductive phase can reduce the productivity of many warm-season vegetables. If temperatures during flowering are below 55 degrees or above 90 degrees, the pollen grains of many warm-season vegetables are damaged and fertilization fails.

For example, tomatoes are pollinated during the day, but fertilization and fruit set take place at night, so it is nighttime temperatures that matter. While the optimum temperature range for tomato growth and flowering is between 72 and 82 degrees Fahrenheit, optimum fruit set requires cooler nighttime temperatures, between 60 and 70 degrees. Fruit set is greatly reduced outside of this temperature range and fails completely if nighttime temperatures are above 92 degrees Fahrenheit. High humidity, which often accompanies high nighttime temperatures in summer, can also reduce tomato fruit set, if pollen grains become too sticky with moisture to make the transfer from male to female flowers.

Many other vegetable crops are sensitive to extreme temperatures during the reproductive phase. Daytime temperatures above 90 degrees Fahrenheit can stop fruit set altogether on bell peppers, especially under dry conditions, and even temperatures in the 80s can decrease yield by half. Daytime temperatures over 86 degrees cause squash, pumpkin and other cucurbit flowers to close, so pollination must be accomplished by bees within a few hours of flowers opening early in the morning. And the cucurbits—squash, cucumbers and pumpkins—that have separate male and female flowers on the same plant tend to produce all male flowers when daytime temperatures are 90 degrees Fahrenheit or above.

The increased summer heat may be the reason Ken has had to adjust the seasonal planting schedule for tomatoes. "I have worked for years with essentially the same timing of my tomato plantings. We do one hoophouse planting and then four in the field. That's intended to give us tomatoes to harvest from June until mid-October. What I've noticed in the last four or five years is our late-planted tomatoes that we have for years targeted with putting in the field around the twentieth of June, that planting is too early. Those tomatoes seem to be growing faster and ripening earlier, whereas we always wanted that last planting in tomatoes to begin being ready for harvest in early September. So we're starting to plant them later than we ever have before."

Like many other growers around the country, Ken has had to adjust his crop mix and planting plans to adapt to warmer and more variable spring and fall temperatures, though the length of the growing season does not seem to have changed in his area. He has found opportunity in season extension and has been successful in increasing cool-season crop production on the farm by expanding greenhouse and hoophouse space. But more variable temperatures in spring and fall, plus falling market prices led him to drop one of his major crops. "For about twenty years, lettuce was our main crop. We used to grow it for the wholesale market. We don't grow much lettuce anymore. It was always susceptible to hot spells early in the spring season, or too much rain, or too early a cold spell in the fall and so forth, whereas other crops are not nearly as sensitive to that kind of variability."

The well-documented earlier arrival of downy mildew, a devastating disease of melons and cucurbits, has required Ken to adjust his plantings of crops like cantaloupe and winter squash. Downy mildew spends the winter down south in Florida and moves up the east coast as summer temperatures increase. "It used to be that downy mildew would appear in eastern North Carolina in early August and then move westward. We could safely grow susceptible crops up until sometime in August, and then those diseases would come. In the last three or four years, downy mildew has started appearing in North Carolina in June. In response to that, we shifted our plantings of susceptible crops earlier by at least a month because if we plant it later, it all dies before it matures."

Ken is fairly confident that, under current climate conditions, he has the resources he needs to continue to farm successfully. "There's really a lot of variability here in central North Carolina, probably more so than in a lot of other parts of the country. We kind of take it for granted that there's going to be wet periods and dry periods and unusual hot and cold here. We're kind of used to it already. It just seems like climate change will require us, at least in this part of the country, to kind of up our game of adaptability and diversity and so forth. I think the reality is we've got to recognize changes are happening and adapt to them. It's high time that we take that seriously and get on with it."

Ken has served the Southeast for many years as a respected leader in sustainable agriculture and local food systems. In 1993 he was named Carolina Farmer of the Year by the Carolina Farm Stewardship Association, a regional sustainable agriculture organization serving North and South Carolina.

Alex Hitt, Peregrine Farm, Graham, North Carolina

Alex and Betsy Hitt established Peregrine Farm on 26 acres of pastures and woodlands in the Piedmont region of central North Carolina in 1981. Although the Hitts initially started a pick-your-own berry enterprise, they eventually moved into five acres of mixed vegetables and cut flowers to improve profitability and meet local market demands. Since 1991, the farm has supported them without the need for off-farm employment, and they also bring in two full-time employees during the growing season. They have never participated in any government program supporting agricultural producers.

Today, Alex and Betsy grow about four acres of vegetables and cut flowers in rotation with a diverse mix of warm- and cool-season cover crops. Production takes place on drip-irrigated raised beds in the open or under about an acre of high tunnels and hoophouses. Extremely diverse crop rotations and intensive cover cropping are key management strategies at Peregrine Farm, with more than two hundred crop varieties grown in a given year, plus about half an acre of blueberries. For more than a decade, Alex and Betsy also rotationally grazed about a hundred turkeys through the croplands each year, but they stopped in 2014 when a local processing plant closed. The Hitts sell most of their produce at a twice-weekly farmers' market in Carrboro, about fifteen miles from the farm, and to local restaurants and a co-op grocery store.

In the thirty-three years he has been on the farm, Alex says that changing weather patterns have caused some major shifts in crop management. "Back in the late '80s and early '90s," he recalls, "we had a number of years when it rained like hell, particularly in early spring, and we many times wondered if we were ever going to get anything planted or weeded. This is when we developed our system of raising our beds up in the fall, so they would drain and warm up fast the following spring when the heavy rains would come. We had so many floods in our creek-bottom fields that we finally had to stop using those fields, even though they have the best soil on the farm, because we couldn't afford to lose the crops. But after Hurricane Fran in 1996, the tap turned off."

Since then, weather patterns seem to have shifted significantly, while extremes have become more intense, creating new challenges and some new opportunities. "In the last fifteen years or so, springs have become much drier, and there are more dry periods and longer periods of

FIGURE 5.9. Alex and Betsy Hitt, Peregrine Farm, Graham, North Carolina.
Credit: Kate Medley, Southern Foodways Alliance

drought in the summer," according to Alex. "Summer high temperatures now seem to extend into the late summer and fall, so summer is longer than it used to be and much drier. But fall weather is also extending longer and is better for growing."

Alex started noticing a decline in production of some crops, particularly tomatoes, as summer temperatures increased and drought became more common. He explains, "In 2012, high temperatures were near 100 [degrees Fahrenheit] for more than two weeks in early June. We've had some heat-related pollination problems in tomatoes, squash, beans and cucumbers. Temperatures were just too hot for fruit set." High fall temperatures as the crops mature have also caused some problems, sometimes actually cooking the fruits on the vine. Drought has also interfered with normal plant development, causing time to maturity to become more irregular in growing seasons with more frequent dry periods and droughts.

> We have a creek that runs by our property. Back in the 1700s, they built two mills and mill dams on this creek. You don't put that kind of effort and energy into a creek unless it is a perennial stream that runs all the time, so it seems likely that the creek has always had pretty good flow all through the summer. We have seen it run dry in some historic droughts, like in 2002, but that was a very rare occasion, and the old-timers said they had never seen it run dry. But in the last 6 years, it has run dry every year at some time in the summer between June and September.
>
> — Alex Hitt

Water availability for crop irrigation is now at the top of the list of weather-related concerns at Peregrine. Water comes from two ponds on the farm, both of which are spring fed. But the springs have not run much for some years now, so Alex pumps water out of the creek and into the ponds as a backup. He says that the creek running dry in summer has raised concerns about having enough water to continue to grow crops in the summertime. As he puts it, "If we don't have the water, we can't grow vegetables in summer."

Peregrine is not in a rapidly growing area, so increased competition for groundwater does not seem a likely explanation for the reduction in summer creek flows. "There are no subdivisions or industrial uses, and, thank god, no fracking or anything yet," says Alex, "so I don't see any large users of water. The area is still mostly in woods." He thinks the summer dry-up may be related to a decline in winter precipitation, which has reduced groundwater levels. "We used to get really good, regular, steady winter rains which kept things moist and green," Alex explains, "but for a number of winters now, you can go out and till soil almost anytime you want. It's not soppy wet. Once the trees leaf out and start drawing down the soil moisture, the creek flow really starts to drop." Two or three summers ago, the creek went dry so quickly that Alex walked its length to the headwaters to see if someone was actually pulling water out, but all he found was that none of the springs that feed the creek were running.

The changes in water supply, coupled with higher summer temperatures and more frequent drought, have got Alex and Betsy thinking about ways, both old and new, to reduce summer crop production risks. For example, soil management, always a priority on the farm, has taken on new importance. "Because we are so conscious of ground water and the creek," says Alex, "we're trying as best as we can to build soil organic matter levels in order to improve soil water-holding capacity. We have a sandy loam, so it dries out pretty quickly."

They have also begun to reduce production during peak summer heat (late June to early August) and focus on production during the cooler fall, winter and spring seasons. This shift away from mid-summer production offers a number of advantages, including reduced water needs, less fieldwork in high temperatures and the production of cool-season crops well-adapted to the longer falls and warmer winters. "From 2000 to 2010, we marketed produce from April through about mid-October," Alex explains. "In 2011, we tried some winter marketing, and that worked well enough that we planted a full array of fall and winter vegetables and some flowers to bring to market in 2012 and 2013. It's an exciting new direction for us."

Warmer winters and a lengthening fall season made the shift in production pattern easier, but brought some challenges too. "As we were trying to get fall crops established last year," Alex says, "I realized why we stopped doing that so long ago. The insect pressures and disease pressures are so high in the fall. It is a struggle. But if we can get to October, we're okay as far as the establishment of crops.... After that, we can go all the way to Christmas easily, without any real additional work. And

January and February are much easier than they used to be, because it is warmer."

Another new weather-related challenge is changing crop disease pressures. Downy mildew and powdery mildew seem to be coming in earlier in the year than they used to and more novel diseases are challenging production. "This year [2014]," Alex says, "I've talked to a number of growers who planted winter squash at the normal time, but because the mildews came in so early, they did not get a crop. We fortunately planted ours really early, and we got a good harvest, but if we had waited any later, I'm not sure we would have gotten much. So part of it is earlier arrival of some old diseases, and part if it is new diseases. For instance, this year the downy mildew that has been infecting basil, which we have never had any trouble with—it finally got into our place somehow, and we lost all of our late basil." Alex adds that some diseases that used to cause losses, like bacterial leaf spot on peppers, have not been a problem over the last few years at Peregrine.

More intense extreme weather, in particular more intense wind, has caused significant damage to the farm. "The intensity of the storms is getting bigger," says Alex. "Snow is more, wind is bigger, and weather comes all at once instead of being spread out." One extreme wind event in July of 2012 damaged 90 percent of the high tunnels on the farm. "In the ranking of storms we have weathered over the last three decades, this relatively small thunderstorm stands at number two in intensity and number one in monetary damage," Alex explains. "Of course, Hurricane Fran will (or hopefully will) hold the top spot forever in wind speed, flooding, trees down and length of power outage, but we had no serious damage to any building or equipment from Fran and not too much crop loss. We have seen record rainfall events [ten inches in an hour and subsequent flooding], we have seen the record snowfall [twenty-plus inches], huge ice storms and hailstorms, but most of those just resulted in loss of power. This storm was fast and hard. The big straight-line winds came screaming from the west, and from our experiences in Fran [80-mph winds for hours] and other hurricanes like Isabel [60-mph winds for a long time], we estimate these winds at 65–75 mph, but for only about ten minutes. The rain lasted maybe forty-five minutes, then it was over." Six of the eight high tunnels on the farm sustained major damage because the winds exceeded their design limits, and the suddenness of the storm caught the Hitts unprepared. The losses from this storm have got Alex and Betsy looking into how they can manage tree lines for improved wind protection in the future.

Even though summers in the Piedmont have been a bit cooler since 2012, Alex and Betsy plan to continue their efforts to enhance the adaptive capacity of their farm to changing climate conditions. "I think we have been lulled into a little calmness here these last two years," says Alex, "at least on the heat end. I keep waiting for it to come screaming back." Even with the retreat from mid-summer production, securing water for crop production remains a top priority. "We continue to go back to thinking about water capture," says Alex. "Are there any other ways that we can control water before it leaves the farm so that we can have it to use? Some of that has to do with windbreaks so we have lower evapotranspiration. We also have places for more ponds so that we can store all the water that does fall on our farm." They also continue to select for crop cultivars that are well adapted to their farm conditions and believe that protected growing space—under row covers and in hoophouses and high tunnels—will become even more important for successful production as climate change effects intensify in coming years.

Both Alex and Betsy are longtime, active participants in their community and regularly participate in and lead workshops at sustainable agriculture and regional food conferences and events throughout the Southeast. Both have served on the board of the Carrboro Farmers' Market. Peregrine Farm was profiled as one of sixty model US sustainable farms and ranches in the USDA-SARE publication *The New American Farmer: Profiles of Agricultural Innovation,* and Alex and Betsy were nationally recognized for their innovative sustainable management with the 2006 Patrick Madden Award from the USDA Sustainable Agriculture Research and Education Program. In 2010, Peregrine Farm was profiled in the NAS publication *Toward Sustainable Agricultural Systems in the 21st Century.*

Challenges to the Adaptive Capacity of Sustainable Vegetable Production

The sustainable vegetable growers featured in this book are responding successfully to the challenges of farming in a changing climate. Many growers report challenges to farm management from more variable weather and frequent and intense weather extremes; however, many have also found opportunity in longer growing seasons. Growers in the West typically mention challenges related to long-term drought and water supply, while those in the East tend to report excessive rainfall, flooding, heat wave, drought and novel disease pressures.

Sustainable vegetable growers are upbeat about their ability to sustain profitable production on their farms under current climate conditions, but identify a number of barriers to the adaptive capacity of sustainable vegetable production systems to climate change effects. These barriers are related to the availability of water, novel disease pressures, proposed food safety regulations, limited federal risk management support and domestic fair trade issues.

Endnotes

1. California Water Action Plan. 2014. resources.ca.gov/california_water_action_plan/
2. The Food Safety Modernization Act gives the Food and Drug Administration broad new powers to prevent food safety problems, detect and respond to food safety issues and improve the safety of imported foods. The regulations focus on food safety risks created by microbial contamination in fresh and processed vegetables and fruits but do not cover meat, poultry or milk or food safety risks from pesticide use or antibiotic resistance. There is widespread concern that the new rules, as proposed, unfairly burden small and mid-scale farms, create barriers to sustainable and organic farming, and ultimately will reduce the availability of fresh, local, sustainable food. The National Sustainable Agriculture Coalition is an excellent source of current information about federal policy and sustainable agriculture in the US. The National Sustainable Agriculture Coalition. 2014. The Food Safety Modernization Act. sustainableagriculture.net/fsma/overview-and-background/
3. Historically, only commodity crops have been eligible for federally subsidized insurance, but the 2014 Farm Bill mandated that crop insurance be

extended to fruit and vegetable producers to reduce climate risk, promote crop diversity and encourage increased production of fresh fruits and vegetables for regional markets. The Whole-Farm Revenue Protection program will provide flexible coverage options of up to 85 percent of loss for specialty crop, organic and diversified crop producers. The new program is designed to meet the needs of farms selling two to five commodities. The program will be available in some counties starting in 2015 and is expected to be available to all specialty crop producers over the next several years.

4. The Agricultural Justice Project is working to ensure that farmers, farm workers and all food workers earn a living wage with full benefits for a forty-hour work week through the use of their Food Justice Certified label. agriculturaljusticeproject.org

CHAPTER 6

Fruits and Nuts

The sustainable fruit and nut growers featured in this book are located in nearly every major region of the country. They represent a variety of models of successful diversified tree crop production systems, ranging in scale from one hundred to five hundred acres and offering their experiences managing fruit and nut production at one location that range from twenty-eight to more than forty years. Many use certified organic practices, and all view soil quality, biodiversity and orientation with value-added direct and direct-wholesale markets as key production risk management strategies. Some of these award-winning tree crop growers report changes in weather conditions that are outside their experience of normal variability and extremes, but others do not.

Increased variability in spring weather and more frequent extreme events have complicated fruit production in many regions and water—too much and not enough—is a key concern. Growers in the West have some concerns about their water supply as warmer winters, drought and higher summer temperatures reduce water supplies and increase competition for water. Some eastern growers are facing increasing challenges created by more frequent heavy rainfalls.

The Northwest

Kole Tonnemaker, Tonnemaker Hill Farm, Royal City, Washington

Tonnemaker Hill Farm is on the north slope of the Frenchman Hills near Royal City, Washington, a semi-arid region in the central part of the state. Brothers Kole and Kurt Tonnemaker are the third generation to own and operate the 126-acre farm, established in 1962 as one of the original farm units in the Columbia Basin Irrigation Project. The farm is home to 60 acres of orchard, 20 acres of vegetables and 40 acres of hay, all irrigated and certified organic.

Kole, his wife Sonia and their son Luke manage the production of more than four hundred varieties of apples, peaches, pears and other fruits and vegetables. Kurt manages direct sales to restaurants, a CSA and as many as fourteen weekly farmers' markets in the Pullman–Moscow area, Seattle and along the I-5 corridor. Farm products are also marketed year-round at the Tonnemaker Farm's popular retail store in Royal City. Besides fresh fruits and vegetables, the farm sells a variety of value-added

FIGURE 6.1. Kole, Sonia and Luke Tonnemaker, Tonnemaker Hill Farm, Royal City, Washington. Credit: Kole Tonnemaker

products made from farm-grown produce including apple cider, fruit leathers, dried and frozen fruit and dried peppers.

When Kole took over the management of the farm in 1981, like many other farms in the region, Tonnemaker Hill specialized in apples, cherries and pears grown for wholesale commodity markets. Faced with falling wholesale prices as the fruit industry transitioned into corporate ownership during the 1980s, Kole began to diversify into higher-value direct markets to maintain farm profitability. In 1992, Kole's brother Kurt took on marketing and sales for the farm full-time and began to expand direct market sales. He also encouraged Kole to consider using organic practices as a way to add value to Tonnemaker Hill products. Kole started the transition to certified organic production in 1997, and a decade later, all of the cropland on the farm was USDA-certified organic.

> Sometimes the challenges of this year keep you from noticing the long-term trend. I think that could be true in our case. I mean, truthfully, looking at the long range, I don't know that we've really seen a big climate change effect here yet. It seems like our struggle is the variation from one year to the next. It just seems like that has overshadowed any long-term effect. Every year seems to be so different.
>
> — Kole Tonnemaker

Kole can't say that he's really noticed any change in weather variability over the thirty-plus years he has been managing the farm; however, he believes that the fall season has lengthened. "We figure on our farm, if we can get past the tenth of October without a killing frost, that's a great year. It's been about ten years since we had a killing frost in September." Because winter kill and spring freezes and frosts are standard risks in tree fruit crops, Kole wonders if year-to-year variability has made it harder for him to see any clear trend over the years.

Another factor influencing his perception of weather patterns may be the farm's north-facing aspect, which makes the production of stone fruits particularly challenging. "The cooler conditions on these slopes just increase the winter kill and spring frosts challenges," he explains. "That's been an ongoing issue for a long time. We grow stone fruits, which are more sensitive to variable weather in the winter and spring. They bloom earlier, and they're more susceptible to winter kill, so that's always something that's on our mind. Once the stone fruits—cherries, peaches and nectarines—have broken dormancy and started to lose their cold hardiness, they cannot reacquire it. They're very vulnerable to temperature variability. If you get a warm week in January and they start to lose dormancy, and then all of a sudden you get a cold spell in February, damage to the fruit bud is a big concern. The apples and pears, which are pome fruits, are less sensitive because they can reacquire cold hardiness if temperatures fall again after a short warm period in winter or spring."

Kole remembers his first decade or so as farm manager went pretty smoothly weather-wise, although major crop losses in 1985 and then again in 1991 and 1992 from extreme weather got his attention. "We're growing perennial crops that are susceptible to just being totally wiped out by freezing weather," he explains. "In 1985, we had a terrible freeze in early November. I mean, it was 25 below zero in the first half of November. The trees were not ready for winter yet, and it killed all the fruit buds. We had lost all our stone fruit crops for the 1986 season, and we weren't even in 1986 yet. And then, in both 1991 and '92, we lost most of the fruit crops again." Abnormally low temperatures in December 1990 caused massive winter kill that devastated the 1991 crop. One of the warmest winters and earliest blooms on record in 1991–92 set the orchard up for total crop loss from the most devastating spring frost ever experienced in the region. "We just thought, oh my goodness, we have got to have something to plant that's not a perennial. We need something we can harvest when we don't have the fruit crops. Also, we started getting insurance on the most risky of the crops after the 1992 frost."

Because Kole had already begun to diversify markets and Kurt was able to take on direct marketing full-time, the decision to diversify into vegetable crops in the 1992 season was relatively easy. Looking back, Kole appreciates the complementary nature of the vegetable enterprise to overall farm performance. The short-season annual vegetable crops allow them to change planting date, crop mix and production volume in response to variations in the fruit harvest and seasonal weather conditions. "We've seen these extreme variations, and that was the killer for us. With fruit production, you need a constant. That's why we went to vegetables. You have some room to modify the production of the annual crops to fit your needs and the season. For example, although we like to plant vegetables in the field starting the first of May, if it's cold, we can just wait."

Access to ample water supplies, industry changes in the 1980s and the back-to-back crop failures in 1991 and '92 that pushed Kole to diversify production and marketing all served to enhance the adaptive capacity of Tonnemaker Hill Farm. Although access to water has not been an issue in his region, Kole has some concerns about future water supplies. "Sixty years ago, this region was just desert. Nobody lived out here. Nobody. The land that our farm is on used to be a massive cattle ranch that went fifty miles one way and sixty miles another way. Now it's a big fruit-growing area. We all get our water out of the Columbia River. Right now, we basically have all the water we need. But already in the Northwest now, there is a struggle to make sure there's enough runoff for the salmon to migrate. One thing they talk about here is if we do get this global warming, it is possible that the Pacific Northwest will get drier. There's been talk about that."

In 2013, Kole and Sonia Tonnemaker were named Farmers of the Year by the Tilth Producers of Washington for their leadership and innovation in "ecologically sound, economically viable and socially equitable farming practices that improve the health of our communities and natural environment."[1]

6.1. Evaporative Cooling and Water Use in the Pacific Northwest

New trends in industrial fruit production in the Pacific Northwest may be on a collision course with the climate change. Projected increases in summer temperatures and more limited water supplies will challenge the current industry trend toward intensive high-density plantings of dwarf trees with a narrow, more open canopy. The dwarf trees can be more profitable because they mature more rapidly and reduce labor costs because they are easier to prune and harvest; however, these intensive orchards are expensive to establish and are more vulnerable to crop losses than more traditional orchard plantings. Less canopy cover allows more direct sunlight on the fruit, which enhances the development of the red colors that are highly valued in commodity markets; however, it also increases the risk of damage to the fruit during periods of intense sunlight or high temperatures.

The industry solution to reduce the risk of heat damage is to cool the fruit with an overcrop sprinkling system—an irrigation system that sprays water over the orchard canopy, so that temperatures are reduced by evaporation. Although sprinkling systems protect fruit from high temperatures and intense sunlight, they promote foliar diseases and can complicate the management of other work in the orchard such as nutrient management, crop irrigation and pesticide spraying programs. Sprinkling systems also use a lot of water, typically 25 to 40 percent more than normal irrigation requirements.[2]

Researchers at the Washington State University Tree Fruit Research and Extension Center in Wenatchee are testing overhead netting popular in Europe as a water saving alternative to overcrop sprinkling.[3] This netting could substantially reduce water use for evaporative cooling and offers growers other benefits like protection from hail and bird damage. The netting may also offer worker safety benefits by reducing conditions in the orchard that increase the risk of skin cancer.

The Southwest

Steve Ela, Ela Family Farms, Hotchkiss, Colorado

As a fourth-generation fruit grower on the western slope of the Colorado Rocky Mountains, Steve Ela is proud to carry on a family tradition of innovative orcharding more than a century in the making. Ela Family Farms is a hundred-acre farm with eighty-five acres planted in twenty-three varieties of apples and twenty-nine varieties of certified organic pears, peaches, cherries, plums and tomatoes. Located near Hotchkiss, it is in the "frost-free" region known for having the best conditions for fruit production in Colorado: 300 days of sunshine, low humidity, ample high-quality water, warm days and cool nights and a relatively long frost-free period.

Steve and his parents work on the farm with the help of four employees year-round. They also employ up to eighteen people during the height of the growing season. Ela Family Farms produces about 1.5 million pounds of fruit each year, virtually all of it distributed in Colorado through direct markets as fresh fruit or value-added products such as applesauce, fruit butters, jams and cider. Farm products are sold through the Internet and a CSA, at farmers' markets all along Colorado's Front Range and to specialty food stores and gourmet restaurants throughout the state.

When Steve returned to take over the management of his family's farm after completing college in 1990, he began thinking about diversifying into direct markets and transitioning to certified organic production to improve profitability and environmental quality. He also replaced the existing furrow irrigation system with more efficient sprinkler and drip irrigation, to reduce water use, and began transitioning to new varieties of fruit trees better suited to organic practices and direct markets. Steve can't say for sure if the weather changes he has noticed are just normal

FIGURE 6.2 Steve Ela and family, Ela Family Farms, Hotchkiss, Colorado. Credit: Steve Ela

variations or a sign of climate change. What he does know is that more variable weather and a lengthening growing season have required him to make some significant changes in production practices to maintain the productivity and profitability of Ela Family Farms.

"The farm is in what was once known as a relatively frost-free area," Steve explains. "Historically, it has been in fruit trees since the 1920s, but in the last decade, we have had some spring frost damage every year now. Five or six of our earliest bloom years have been in the last ten years." He has also seen the fall season lengthen noticeably. "There are some varieties of apples, like Fuji, a late-season apple that ten years ago we weren't sure we could grow here. Now we commonly pick them two weeks before the end of the season."

> I used to say it would be one year in ten we would expect a really bad year, maybe another two or three years we would have some frost. Now I would say we have frost every year. The one-in-ten year with a 10 percent crop, that still holds, but now we're having 50 percent crops many other years. When I say this was a frost-free area, it used to be that growers didn't need wind machines and other frost protection measures and they got though just fine. Now we have the whole place covered with wind machines.
>
> — Steve Ela

Although the lengthening growing season has improved growing conditions for some apple varieties on the farm, production risks have increased, particularly in the last decade. "We're experiencing earlier springs and more variable temperatures in the spring," says Steve. "As an example, in 2013, on April 15 we were at 13 degrees. That is more typical of February or March temperatures. To get below 20 in April is crazy, and we had two nights below that. So it's not just early blooming, but late, abnormally cold temperatures. April and May are the huge frost months for us. Peaches bloom mid-April, apples bloom toward the end of April. Any sub-freezing temperatures during that time are pretty destructive. Spring temperatures control whether or not we have a crop."

Variable spring frosts also create a lot of uncertainty in orchard management, because fruit trees are managed to reduce the number of fruits and to evenly space the fruits on a tree to increase fruit size and quality.

The final crop load—the number of fruit remaining on the tree to mature—determines the season's yield potential. "Not knowing from year to year how much frost damage we are going to have means it's much more difficult to manage crop load," Steve explains. "If you're in an area where you're not going to have much frost, you can prune and thin in the fall with confidence, because there is a low risk of losing additional fruits to frost damage. Now, we never know from one year to the next how much winter and spring damage we are likely to get, so we have to leave a lot more fruit out there during pruning and early thinning. If it turns out we don't have frost in the spring, then we're behind the curve getting it thinned off in the spring. It's this not knowing which way to jump that is really difficult."

Steve has also noticed changes in summer and winter weather that have complicated management over the last decade. Warmer and wetter summers have increased disease management challenges. Over the last several years, the hottest time of year has shifted from early August to June, when temperatures regularly reach 100 degrees Fahrenheit, sometimes day after day. Rainfall patterns also seem to be changing. It used to be common for no measurable rain to fall between the end of May through late August, but now consistent light rains in July and August are common. This rain is not enough to water the trees, but leaves them moist enough to increase the risk of disease damage. And winter low temperatures have become more extreme. Recently, Steve had to purchase propane burners and use his wind machines to try and buffer extreme cold temperatures. "Peaches are very sensitive to cold temperatures in the winter," he says. "We bought our first wind machines in 1991, but I've never run them in the winter until the last two years. We have been below critical temperatures for peaches in the winter the last three years, at some point or several times."

Steve has made other changes on the farm to reduce increased climate risk. He has added more wind machines, makes use of microclimates and is considering adding protected growing space. Steve explains, "We had a couple of wind machines in '91. They cost twenty to thirty thousand dollars a piece, so we didn't buy them all at once. We probably put the last one in about eight years ago to finish covering the whole property. We have a hundred acres, so we have eleven wind machines." The farm is on a hill about three-quarters of a mile long with a number of swales and other landscape features that influence temperature. "I say we live in a frost-free site, but on some cold nights, we can have a four-degree difference across the farm," explains Steve. "I have some ground out there that has historically been planted in trees, but I will not plant trees there now, because it is a cold pocket and the risk is too high. I'm looking at the warmer spots on the farm, and that's where I put my most sensitive crops."

6.2. Frost Protection[4]

The terms "frost" and "freeze" are often used interchangeably, but technically speaking, the word "frost" means the formation of ice crystals on a surface. Plants are injured when temperatures drop below freezing, and ice crystals form in extracellular spaces inside the plant. As this ice forms, water is drawn out of nearby cells, and they become dehydrated, causing injury and eventually death from water stress.

The ability of plants to resist frost damage varies among species and also depends on growth phase and the temperature conditions before the freeze event. The amount of injury depends on the crop's sensitivity to freezing at the time of the event and the length of time the temperature is below the "critical damage" temperature. Tree fruit producers are at high risk for complete crop failure during the spring bloom period, typically February through April in temperate regions, when the trees are in a very sensitive stage of growth.

Fruit growers spend a lot of time, effort and money to prevent crop losses during the highly variable spring weather. They use a variety of practices, most of which increase temperatures just a few degrees, which is often enough to prevent damage to the bloom. Passive frost protection measures are proactive biological and ecological management strategies taken to prevent frost developing in the orchard and reduce the sensitivity of the orchard to frost in an effort to avoid the need for more expensive active measures. These include proper site selection, managing cold air drainage in the orchard, plant variety selection, good plant nutrition and proper pruning. Active frost protection measures are used during the freeze event to try and keep temperatures in the tree canopy above freezing; they include such energy-intensive practices as using heaters, wind machines, helicopters or sprinklers.

FIGURE 6.3. A Wind Machine on Ela Family Farms, Hotchkiss, CO. Steve has purchased 9 wind machines (at $20,000 to $30,000 each) since 1991 to protect his fruit trees from increased variability in spring frosts. Credit: Allie Goldstein and Kirsten Howard, Adaptation Stories

Heaters have been used to protect plants from freezing for more than two thousand years. Originally, open wood fires were set in the orchards to warm the air, but by the mid-1900s, metal orchard heaters, called "smudge pots," were standard practice worldwide. Wind machines began to replace heaters as the preferred method of frost protection by the 1950s. Although they cost more to buy, wind machines are easier and less expensive to operate than heaters and produce much less air pollution. They raise temperatures near the orchard floor by blowing air horizontally, mixing warmer air aloft in a temperature inversion with cooler air near the surface.

Steve is also careful to select frost-tolerant varieties, particularly of peaches. "Within peaches, some varieties are more susceptible than others. When evaluating which peach variety we're going to use, keep or replant, I'm looking at that frost sensitivity. We're certainly finding varieties that are more likely to come through a spring frost than less likely, even though that means we may have to do more thinning." Steve is also considering adding frost protection structures to his cherry orchards, because of increasing risk of frost damage in the crop.

Asked about his confidence in the future, Steve notes that he is still in business in an area where fruit farms have declined by 75 percent over the last twenty years. He puts a lot of that down to his choice of direct markets. "We started changing that in 2000 because of bad economics, and now we direct market 100 percent of our fruit. We've completely changed our business model in twelve years. Fortunately it's worked, we're still here. But we've made a conscious effort not to play in that international or even national commodity market. We have access to a little higher-value market, where we have more control."

The high returns possible with direct markets have buffered the increased production risks the farm has faced over the last decade. Steve notes that direct markets have also opened up new opportunities for him to diversify crops, because his customers are willing, even eager, to try something new. According to Steve, "with the direct marketing, we have a little more control on price, which means we don't have to hit a home run every year to still be viable. I've looked at the marketing as a way to mitigate some of that crop production risk. Can we still make money if we have a half crop versus having to have a full crop every year?"

He goes on to explain some other benefits of selling his crop this way: "Direct marketing provides some additional risk management because it also means we can pick more varieties that maybe aren't suitable for wholesale markets, but maybe have characteristics we can handle in direct markets—for example, a variety that's frost hardy but doesn't ship well. So we can pick and play with some of those varieties that we haven't been able to before." Steve believes that the uniform product requirements of industrial commodity markets increase risks in fruit production because growers are not free to select varieties best adapted to their particular farm conditions.

But there is a downside to direct marketing—it takes a lot of time and some additional skills, and it keeps Steve out of the orchard. "I now spend 50 to 60 percent of my time marketing," he explains, "whereas ten years ago, 80 percent of my time was growing. I have become a worse grower because I have to spend my time marketing. As a farm, that has been a good trade-off. We are doing much better than we did before. But I would rather be a grower than a marketer. I'm a decent marketer, I don't hate it, but I would still rather be a grower. Choosing this marketing avenue that takes a lot more of my time is in part about risk management, which is in part about weather."

Like many growers in the Southwest, Steve has grave concerns about the future of his farm's water supply, which is renewed each year by snowpack meltwaters. "Water management is always a concern for us because we're dependent on irrigation. We're going to look at the snowpack each year to determine how our water management might have to change. Every year it is different. Our average rainfall here is ten inches, and it does not necessarily fall in the summer when we need it. I've had people say to me that, with climate change, it gets warmer, and you guys will be set. No, climate change is more variable, which doesn't help us, and if it's warmer, we have less snow. We're absolutely dependent on irrigation water in the summer. And if that regularly becomes less, it will definitely put a crimp on what we can do."

Steve has leased a neighboring farm purely as insurance against drought. "On this farm we're on, we have adequate water rights in average years," Steve explains. "In dry years we're short, so we lease a neighboring farm that is largely fallow right now, mostly for the water, just because it keeps me from getting more gray hair. It's an insurance policy, that's what it is. And if other water rights that we can access come up for sale, I'm going to be right in there trying to buy them. Water in the West has always been competitive. If it decreases, and especially if we continue to have population growth, there's going to be greater and greater pressures on that water for domestic use. It's going to get ugly."

Because tree fruits are long-lived, Steve is hopeful, but concerned about the nature of the climate risks facing the farm. "We're investing a lot of money into planting new trees. It costs somewhere around eight to twelve thousand dollars in the first year to plant a new acre of trees, and it's a ten- or twelve-year payback period if we do everything right.

So anytime you put more risk in that equation, it's scary. It's a dilemma. You can't really quit planting out of fear, because if you don't renovate, plant and keep moving forward, pretty soon you're going to have a bunch of old trees, with nothing coming up beyond them to support the farm. It's a catch-22 and that is unnerving, and that worries me. It's certainly something I've thought about quite a bit. I feel confident that we have access to some of the best tools and information out there, but does that mean we're going to successfully manage it? I'm not confident of that at all. Ultimately it's going to come down to what is economic and what makes sense."

Steve served as board president of the Organic Farming Research Foundation from 2004 to 2009 and regularly collaborates with University of Colorado researchers on organic fruit production research.

Midwest

Dan Shepherd, Shepherd Farms, Clifton, Missouri

Dan Shepherd helped plant the first fifteen acres of pecan trees at Shepherd Farms near Clifton Hill, in north-central Missouri, when he was fourteen years old. The pecan orchard was just the first of many alternative crops that Jerrell Shepherd, the farm's founder and Dan's father, put into place on the 1,900-acre corn, soybean and wheat farm he purchased in the late 1960s in an effort to improve profitability through diversification into high-value specialty crops. Dan continued his father's tradition of innovation when he took on full-time management of Shepherd Farms in 1985 by adding buffalo and gamma grass to the annual grains and pecans grown on the farm. Dan integrated all the crops and livestock through an innovative agroforestry system featuring alley cropping and management-intensive grazing.

FIGURE 6.4. Dan Shepherd, Shepherd Farms, Clifton, MO. Credit: Tim Baker, Univ. Missouri Cooperative Extension

Although the buffalo and gamma grass are now gone and most of the 4,000-acre farm is once again in an annual grain rotation, Dan still manages about 300 acres of mature pecan orchards. Pecans are processed on the farm, in a purpose-built facility that cracks, shells and packages them for direct market sales. Dan markets his pecans and other products through the Internet and an on-farm store open from October through December each year.

> I've been out here for 40 some years, it's really hard for me to notice any changes in weather patterns. As for the moisture, the drought and the frost and the freeze, we've always had those problems. I really can't see a whole lot of change, even in 40 years.
>
> — Dan Shepherd

Like tree fruits, the production of tree nuts is complicated by variable weather during periods of temperature and moisture sensitivity in the annual life cycle of the plant. Over the years, Dan has learned how to produce a profitable crop of pecans despite the highly variable weather that is normal in his region. Pecans are sensitive to cold weather and frosts during the spring bloom, which typically occurs in late April. Drought in the summer and fall during the period of nut fill can cause small and misshapen nut meats. And although pecan trees are considered flood-tolerant, flooding anytime except during winter, when they are dormant, can stress the trees and reduce nut yields.

Dan can't say that he has perceived any change in the weather over the forty-plus years he has lived at Shepherd Farms. Through the years, the pecan bloom has been hit by frost pretty regularly, about once every four years. Sometimes this actually improves yields because a mild frost will reduce the nut load just enough to improve yield and quality. The diversity of pecan varieties also helps reduce the risk of spring freeze damage. In most years, at least some of the eight varieties grown in the orchard escape frost damage completely, and rarely have any of them suffered a total loss due to spring freezes or frosts.

The Shepherd Farms' orchards are not irrigated, so drought in the late summer and early fall has been a challenge at times; however, Dan says that summer temperatures or the frequency of heat waves or droughts has not changed noticeably at the farm over the last forty years. "Variability in precipitation affects the pecans more than anything else," he explains; "I need a rain in August, and if I don't get it, the pecans really suffer." But pecan trees are tough once they've had some time to get established, especially to temperature extremes: "In the wintertime, I don't care what the weather brings," Dan says. "These pecan trees, the central and northern varieties that I grow, are pretty tough. The coldest day we've had so far this winter [2013] was 17 below, and I'm not worried about anything that's three years or older. In summertime the heat just doesn't affect them, it gets up to 100 or 103 degrees, it's no big deal, they're made to take that, temperature-wise. Moisture-wise, they can take a flood in the wintertime, it really doesn't hurt them, but any other time, a flood does. And dry weather hurts them in the summer and fall."

Pecans are native to the bottomlands of the Mississippi River basin and are well adapted to the wet conditions and recurring floods. The Shepherd orchard is planted in fertile floodplain soils along the east fork of the Chariton River. The river floods quite often, but the Shepherds built a levee in the 1970s that protects the orchard. "I've got a ten-mile levee on this farm," Dan explains, "and that's my lifeblood. My father put levees in, and we built them all ourselves; they're private levees, and we treat them with great respect. We keep them mowed; it's our main farm road, and we keep them up. That levee system is my whole lifeblood."

"I do have a problem with flooding in my part of the country," Dan goes on to say. "I'm in the river country up here in Missouri, and I'm on a river that does flood regularly. I've seen floods in every month of the year. Some of the worst floods we've had were in spring, when the ground's still frozen. You get a bunch of snow on the ground, the frozen ground, and you get an inch of rain on top, you can have a pretty good flood. Even though we have a levee on the farm, and it's probably one of the best levees in the country, it still can be topped like it was last year [2013]."

Shepherd Farms experienced unprecedented flooding and drought in 2013. "One of the worst floods we've ever had on the farm was in April 2013, and one of the worst droughts we've ever seen came that summer," Dan says. "We got them both in the same year. Flooding really hurt the pecan trees in the spring, being underwater for a week or two, and that really set them back. Then turn right around and June 23 was the last rain we had until somewhere up in the middle of October, so that hurt the fill on the pecans." Thinking back on that year, Dan says, "We got hit with water both ways in 2013. Too much water and not enough and at the wrong time of the year, that is tough."

Although Dan has not noticed any clear trends in changing weather patterns, the last decade or so has included several unusually extreme weather events on the farm. He experienced total crop loss from a spring freeze for the first time in 2007, and the levee was breached for the first time since it was built in 1970 by a flood in July 2008, and then again in 2013. Dan can't remember another time in the forty years he's been on the farm when there were so many dry summers in a row, like those in 2011, '12 and '13. But Dan sees these events, though unusual, as just part of life in Missouri. "I think a lot of this is just typical of weather in my region," he says. "We've got the Gulf of Mexico just south of us; it pumps a lot of moisture up. We have the big mountains in Colorado out there to stop those systems and dry them out before they come. We've got Canada up north that can drop a lot of cold air on us. Our weather is so variable that it's hard to get a grasp on any changes that might be coming down the pike."

Dan learned a lot from his father about using weather forecasts in farm planning and fieldwork scheduling. Dan explains, "My father was really an excellent weather forecaster. He was mainly in the radio business. He loved farming, but he also ran a bunch of radio stations. He had the first Doppler radar in Missouri in a radio station, right here in a little town nearby."

Dan also has an interest in climatology and has read widely on the subject of climate history. "Do we have climate change? Sure we do!" he says. "I can remember back in the mid '70s to the early '80s, we were going into a little ice age. Then we were supposed to be burning up, and they're going to call it global warming. And now they call it climate change. I'm enough of a climate history buff to know that we go through these cycles. We have twenty-year cycles, we have two-hundred-year cycles. When the Thames River froze up in the little ice age and when we had the terrible winters of the early 1900s. A lady here in Clifton Hill said that she remembered it snowing on June 6 in 1911, but that was in 1911, and we had global freezing then. Those are weather cycles."

Thinking about the future, Dan expressed concern about the drought situation in California. "They raise so much of our food that we're definitely in trouble there. I was just reading something the other day about some of the droughts they've had in California, and it wasn't too long ago, like 500 to 650 AD, that California had a 150-year drought. And back before that, around the Year Zero, there was a 180-year drought. What are we going to do when we get a 180-year drought in California?"

Dan Shepherd's work developing gamma grass as a native forage crop and his innovative agroforestry system integrating grains, nuts, forages, buffalo and seed crops has been nationally recognized by the USDA's Agriculture Research Service, the National Agroforestry Center and the Center for Agroforestry at the University of Missouri. Shepherd Farms is one of sixty farms and ranches selected for the USDA-SARE publication *The New American Farmer: Profiles of Agricultural Innovation.*

Jim Koan, Almar Farm and Orchards, Flushing, Michigan

Jim Koan has been growing apples at Almar Farm and Orchards in eastern Michigan near Flushing for more than forty years. Although the soils in eastern Michigan are heavier and the climate more variable than

FIGURE 6.5. Jim Koan, Almar Farm and Orchards, Flushing, Michigan. Credit: Richard Lehnert, Good Fruit Grower

the ideal fruit-growing conditions found in western Michigan, Almar is Jim's home, and he wanted to continue the Koan tradition of growing apples there. When he took over the family business in the mid-eighties, Jim grew apples using industrial methods like his father before him. But after a decade managing the farm, he became interested in integrated pest management (IPM) as a way to cut costs and reduce environmental impacts. Early success with IPM encouraged him to make the transition into certified organic production.

Today Jim, his wife Karen and three of their five children work together on Almar's 500 acres, producing thirty varieties of organic apples in a 150-acre orchard as well as pumpkins, corn, soybeans, wheat, barley and pasture. About 30 acres of apples are intensively managed for fresh market sales, while the rest are processed on-farm into hard cider and other apple products. About 150 pasture-raised hogs are farrowed and finished each year on the farm. Jim uses a Swedish sandwich system to reduce soil erosion and enhance soil quality in his orchards and manages native wildflower planting to encourage beneficial insects, reduce pests and improve apple pollination.

> Two years ago, the whole state of Michigan had a ten percent crop of apples. Worst freeze since 1945, I believe. Then this last year, again the same thing occurred, and we had another significant freeze. Two years in a row of those extreme freezes have never been seen before in my lifetime or even by fruit growers who started growing in the thirties and forties. Spring frost is getting to be a bigger and bigger problem.
>
> — Jim Koan

Jim has been recognized over the years for a number of innovative practices on his farm, but most recently, he has received a lot of attention for the successful integration of livestock—pigs and poultry—into his apple production system. Jim pastures heritage-breed pigs in the apple orchards to clean up fallen apples that harbor the plum curculio, a weevil that is one of the most destructive pests of organic apples. The pigs also help to build soil quality and manage weeds. Jim feeds them on apple pomace, the paste left over from pressing apples for cider. Jim direct markets his finished hogs as pasture-raised, apple-finished pork and sells a number of other value-added products from the apples he grows on the farm, including apple cider, apple cider vinegar and an award-winning hard cider that has been made on the farm since the 1850s. The hard ciders are distributed nationally, while fresh apples and the other processed products are sold directly through an on-farm store.

Over the last decade, Jim has noticed a number of weather changes that have increasingly complicated his farm management. Weather extremes are getting more extreme, and disease and insects seem to be getting harder to control. "With the changing weather that we have now," Jim says, "every year the extremes seem to be getting more extreme. If we're going to get rain, we're going to get a lot more rain than usual. We'll get deluges, not the three or four inches of the past, but we'll get five or six inches. Or we'll get snowstorms, or extreme heat, or droughts and so forth. Those are very disruptive to the natural balance of nature, of insects and disease."

Variable spring weather has always been a significant factor in tree fruit production in Michigan and elsewhere, but as Jim explains, the risks have grown. "Recently, the biggest challenge has been the warmer weather in March. Normally we'll get several days in the 60s and then we'll drop down again to some 30-, 40-degree days, while the nights drop to freezing or below freezing. Now, instead of getting a few days of 60, 65, it will be 75, 80 degrees. Well, it doesn't take many days of those kind of temperatures to accelerate our trees waking up, and then we end up with earlier bloom even though we still usually have a few significant freezes in the later half of April or early May. Two years ago [2012], the whole state of Michigan had a 10 percent crop of apples. Worst freeze since 1945, I believe."

Jim wonders if part of the trouble has to do with the Great Lakes not freezing over in winter like they used to. "Michigan has always been a great fruit growing area for cherries and apples and peaches and what not, your perennial crops," Jim explains. "Because the Great Lakes, that ice was a huge cold sink.[5] So in March, when we did get these little warming trends that might wake the trees up, we had that ice all around us to help buffer that. The Great Lakes haven't been freezing over like they have in the past, and therefore we have lost that buffering."

Heavy rainfall and storms are becoming increasingly destructive, according to Jim. "Two years ago [2012], in August, we had the most rain in a hundred years," says Jim. "Broke the hundred-year record, okay? In one night, we got six inches of rain—unheard of. The orchards and everything were all flooded over. The water overflooded the banks for the first time that I can remember and ripped out a bunch of trees and fences. The animals were all running around the farm. It seems like we are getting one snowstorm after another now, followed by extreme temperatures and windchills, and then warming trends. It's not just one event every ten years anymore. It's just going from one extreme to the other, and those changes are extreme within the weather cycle."

In an effort to reduce the risks associated with more variable weather, Jim has made several adjustments in production and marketing. He has added more drainage to his orchard, transitioned to more disease-resistant varieties and diversified his product mix. Jim began installing tile drainage in his orchard about fifteen or twenty years ago. Initially he laid tile drains every fifty feet: "In any new orchards before that, nobody tiled orchards. Now on my new orchards, I tile every twenty-eight feet. It's not just that I'm tiling, but that I'm actually having to get them closer together to get the excess water out of the soil more quickly." Jim is quick to point out that not all growers in his region have had to add additional drainage. His orchards are on heavy soils with poor drainage to begin with, but more extreme rainfall events have made drainage even more important.

Jim says that other fruit growers in eastern Michigan have adapted to more variable spring weather by adding wind machines for frost protection, and many are abandoning high-risk areas. Jim explains, "Almost all the apple growers in Michigan that can afford it have bought wind machines to protect against frost. If they had two wind machines, they bought two more for other sites where they weren't needed before. They're also looking at replacing fruit crops with grain crops in the poorer sites and only using the very best sites for fruit crops because the input costs for fruit production are so high today that you can't afford not to get a full crop."

Jim used to grow sixteen different kinds of fruits, but today he grows only apples. Increases in production costs, weather-related risks and changing consumer preferences have all played a part in his decision to reduce the diversity of fruit types at Almar Farm and Orchards. "When I first started growing thirty years ago," Jim explains, "it was nothing to sell two hundred bushels of peaches in a couple of days from the farm store. A housewife would come out and buy a bushel or two of peaches and then take them home and ripen them and can them and then two or three days later come back and get another batch, and come back a third time maybe three weeks after that and still get another half bushel to eat out of hand and maybe make some peach pies and cobblers and whatnot. But now, people can go to the store and buy fresh peaches to eat on the table, put in their fruit bowl and eat almost year-round. Those peaches are going to be from Chile or whatever, but at least they can buy them. So they don't can them anymore. People quit canning pears, same thing with peaches. So I don't grow peaches and I don't grow pears. Now I don't sell two hundred bushels of fresh fruit in the whole season, you know?"

While he still maintains a diverse mix of apple varieties in the orchard, Jim says it can be difficult selling them in a market defined by year-round availability of a limited variety of apples. Like Steve Ela, Jim finds direct marketing gives him some flexibility to select apple varieties that are well adapted to the changing climate conditions on his farm. As the weather has become more variable, Jim has transitioned to more disease- and insect-resistant varieties, which sometimes require some consumer education. Jim explains: "I'm planting varieties that are more disease- or insect-resistant, but consumers don't want them because they've been programmed by advertisements to think that Gala is a wonderful apple or Red Delicious or Golden Delicious, Braeburn, Fuji or whatever. More than two thousand commercial apple varieties have been grown in the United States in the past. We're down to just a few varieties now because consumers have been brainwashed that these are what tastes best."

Jim tells the story of how he successfully sold his customers on an apple variety that he is particularly fond of growing, called Gold Rush. "It is a fantastic eating apple, but it is ugly," he says. "It has these big pores in the skin called lentils, and people didn't think it looked good. If I took a Gold Rush apple and put that in a grocery store, the grocery store couldn't sell them because consumers don't know what a Gold Rush is. They know what a McIntosh is, so the Gold Rush is not going to get sold sitting next to a McIntosh. But that's a big problem, because it is extremely difficult to grow an organic McIntosh. We grow four, five thousand bushels a year, but they're extremely difficult to grow. But the Gold Rush is a more sustainable apple. So I put up a sign in our farm store one year, a really big sign that said, 'Gold Rush, the ugliest, best-tasting apple in the world.' That aroused customer curiosity. People went and bought them, and they came back and bought more. Now we've got a really strong Gold Rush consumer demand in our area. I introduce the people to these other varieties, and they love them, and they come back and buy them for eating fresh out of hand. But it's an education. I'm educating a consumer in order to sell these apples. You can't do that for thousands of bushels of fresh apples. You don't have the time to do that, you know?"

Jim has increasingly focused on marketing processed products in an effort to build a sustainable business model. He has sought out products that allow him to avoid competing in international and national commodity markets, as well as allowing him to continue to use sustainable practices like the integration of livestock into his orchard production system. Jim saw the federal food safety regulations like GAP and the FSP as a real threat to his freedom to farm sustainably. So he began thinking about how to transition from fresh products to processed products that would allow him to meet new regulations without having to change production practices. Jim explains, "We raise pigs because they're part of the system. I use them for insect and disease control, and I can sell their meat as a protein source. Almost all of our apples are processed on the farm and made into juice. Fifty percent of that bushel is still food. Even though the juice is taken out and fed to humans, you've got all this other good nutrition left in the pomace. That goes to feed our pigs, and then we use their manure for fertility in the orchard. We work as a team. It looks like with the GAP and new food safety regulations I won't be able to raise livestock on my farm anymore. For a sustainable farm, you have to have an integration of livestock and crops. It's not like CAFO operations where somebody's got a thousand head of swine locked up in the barn and they're pumping corn through them for six months and then selling them."

Jim wanted to find a way to keep the pigs in the production system under the new food safety regulations. He started thinking about processing apples into an alcoholic beverage that would eliminate any food safety concerns. In 2009, just a few years ahead of the boom, Jim developed a line of hard ciders under the JK Scrumpy's label. The new product allowed him to keep pigs in his apple orchard and provided other unexpected benefits as well, like expanding his customer base. "Now I sell interstate all over the United States," Jim says. "I have distributors for my product. It's shelf-stable so I don't have to worry about having to sell it right now."

Jim also learned that a shelf-stable addition to his product mix provided a buffer to weather variability and extremes. "In 2012, as an example," he explains, "we had only a 10 percent crop of apples. I had half a million dollars invested in those apples. That was not as big an issue for me as it would have been if we hadn't had JK Scrumpy's, because the year before that I had had a huge crop. I had fermented a whole bunch of those apples, and they were just sitting there on the farm, in the bank, so to speak. I still had a non-perishable profit from the year before, so 2012 didn't disrupt my cash flow too much. I can walk away comfortably saying that I actually made a profit in 2012, not only because of the surplus I had stockpiled from 2011, but also because I was forced to think out of the box and do things differently. I really came out ahead of the game."

Jim appreciates the opportunities that recent weather challenges have created for his business. He says they have forced him to think out of the box, anticipate what could go wrong and plan for the worst-case scenario. He has focused a lot of attention on developing a business that is robust to what he views as both political as well as climate risks. Thinking about the future, Jim is confident he can handle the biological challenges, but he is concerned about political and regulatory challenges to sustainable agriculture and local food production. "I'm better prepared than people who just go along thinking everything is business as usual," says Jim. "You could say I have developed a sustainable business plan that protects me from both political and weather extremes."

Jim was the first apple grower in Michigan to transition to certified organic production. He has been actively involved in the leadership of many sustainable agriculture and organic farming organizations over the years and is a longtime collaborator in on-farm experiments with Michigan State University faculty and staff. In 2013, Jim's long years of dedication toward the improvement of the Michigan fruit industry was recognized with a Distinguished Service Award from the Michigan State Horticultural Society.

Northeast

Jonathan Bishop, Bishop's Orchards, Guilford, Connecticut

Effective adaptation to changing market conditions has been a hallmark of Bishop's Orchards, a 140-year-old farm located near Guilford, Connecticut. Through six generations, the Bishop farm has evolved from a small general farm peddling ice, milk, fruits and vegetables door to door in the local community, to a wholesale grower of fruits and vegetables supplying regional markets, to a thriving retail market offering a diverse line of fresh and processed products, many produced on the farm.

Jonathan and Keith Bishop are cousins, fifth-generation co-owners and managers of Bishop's Orchards and related businesses. Jonathan is responsible for production, harvesting and warehousing of all crops on the farm, including disease and insect control, integrated pest management (IPM) and the management of farm equipment. Keith is responsible for retail marketing, sales and management of the family business, and is also Bishop's winemaker.

FIGURE 6.6 Jonathan Bishop, Bishop's Orchards, Guilford, Connecticut. Credit: Jonathan Bishop

While apples are a focus of production on the 320-acre farm, Jonathan also manages a diverse mix of vegetable, berry and flower crops for direct sales through an on-farm, full-service retail market and bakery, a winery, a pick-your-own operation and a CSA. Bishop's was an early innovator of IPM methods for fruit production in Connecticut. Jonathan has reduced the use of pesticides on the farm by up to eighty percent through a program featuring scouting, fumigant cover crops, trap crops, agroforestry and other practices that serve to increase biodiversity and reduce pest pressures. The farm and associated packing/cider operation at Bishop's Orchards employs a full-time staff of fifteen and adds as many as thirty seasonal employees during the growing season, while the retail side of the business employs about fifty-five people year-round with an additional sixty seasonal staff.

> We always talk in the course of a year about how the weather seems one way or another, how it's different from normal. I think it also gets to the point over time of not really knowing what normal is. I can remember unusually warm spells and cold spells from when I was a kid. I think what may color my responses somewhat is that Gilford is a shoreline community. Our orchards, many of our orchards, are within a few miles of the coast. So we get a very moderating influence from Long Island Sound.
>
> — Jonathan Bishop

When Jonathan thinks back over the thirty-five years he has been managing production at Bishop's Orchards, several long-term production challenges come to mind. Changes in pesticides, novel pests, insects and disease management and wildlife—particularly deer and voles—have been continually challenging. "Most of the complication on the insect and disease side," Jonathan explains, "is changing chemistries, the phasing out of the organophosphates and some of the longer residual fungicides, the pest-specific nature of the replacements, and some introduced species. The spotted wing drosophila [fruit fly] has become a huge pest for small fruit growers and the brown marmorated stink bug is another one that, knock on wood, we haven't had to deal with yet. It's another one that's out there. These recent pest introductions have happened, I think, as a result of global trade, the nature of trade these days. We've maybe let down the guard a little bit over the years, and the focus has shifted toward trying to find terrorists and bombers and not concentrating so much on some of these other imports that can have major impacts on agriculture."

Weather is always a challenge in fruit and vegetable production, and that has also been true at Bishop's. Like many fruit growers, Jonathan has continuing challenges with variable spring weather, summer drought and periods of moisture that encourage plant diseases. He thinks that dry periods might be the biggest challenge because of all the extra work involved in watering. "On a lot of our small fruit crops, we have trickle irrigation in place," he says. "With the tree fruits and some of the vegetable crops, you get involved in moving pipe around and getting water to the pipes. The dry periods are difficult to deal with in that regard."

Jonathan can't say that he has seen any kind of changing trends in weather. There have always been extreme events through the years, and he doesn't think these have increased in frequency or intensity during his lifetime. He can recall some extreme weather events throughout the years. "For instance, we just went through a pretty cold spell with the Polar vortex [in 2014]. Yet I can remember in 1981 we lost our peaches from three days of minus 12 temperatures. We haven't had that kind of cold since then. We had a really warm February in 1976, the apple buds actually started to swell, and then it dropped back to normal winter temperatures, and some varieties were 100 percent loss that year. The earliest season I can ever remember was in 2012. We started five weeks earlier than normal, but it was followed by last year [2013], which was a fairly late season for us. Of course, there was the Halloween snowstorm in 2010. It's hard to say that there is a trend even in the variations because there's been some pretty big swings going back thirty-some years. Like I said, I've seen extremes, but I haven't seen an increase in the extremes."

Jonathan thinks that the farm's location on the coast of Long Island Sound may have provided some buffer against weather extremes. "The sound may be moderating the absolute cold temperatures in the winter and the hot temperatures in summer. Growers inland often face much bigger issues with spring cold temperatures or frost than we do. That maybe part of why our experience may be a little different from what other people might have noticed." Over the years, Jonathan has learned to be prepared for whatever the weather might bring. "Every season we plan for the quote-unquote normal situation," he says. "We're prepared for reacting to unusual events. If we had an unusually heavy rain and we needed to reapply a protectant to a crop or something, we just figure out what we're going to need to do in terms of having the systems ready to go when we need them. I guess we just try to be prepared for anything."

Although the last thirty years have brought a lot of changes to Bishop's, most have been driven by marketing considerations, not changes in weather, Jonathan explains. "There are so many factors other than weather that are driving crop choices. We've been moving very steadily away from apples, which used to be our biggest crop by far. Apples tied us to wholesaling. Since then, over time we've been using alternative marketing methods that are pick-your-own, through our own retail or the CSA. We have been trying to adjust our mix of crops to match our production to our retail needs. We have been expanding into other crops like peaches, small fruits and a number of vegetable crops. If one thing doesn't work out one year, it's better the next. We've always looked at our diversity as our insurance."

Jonathan appreciates the benefits to risk management provided by diversity, even within just one crop. "It's always interesting—even within a single crop like apples, there will be a year when one particular variety is just outstanding and the quality of another one is just not what you'd hope it would be. We're always looking at our diversification and adding different things to the mix, trying them out, sort of move it around and doing a little bit of our own research and development in-house to find stuff that hits a niche that we want to try to hit."

A number of severe storm events over the last few years confirmed the benefits of scale, experience and crop diversity. Jonathan explains, "Because we're a fairly good-size farm for our area and we're pretty diversified, when we do get a bad weather condition, something that might drive another farmer to have a bad crop typically has less of an effect on us." One example is with the CSA. "There are quite a few CSAs starting up in our area. There's a small farm not too far from us who suffered pretty badly a year or two ago. And that's okay, their CSA members understood that it was a bad year. But the following year when they're looking to be in a CSA, and they have a choice between a CSA that's supplied people with something all season long or a CSA that basically gave up in July.... We have a lot of people that have previously been with somebody else who signed up with us because we have more capacity to make sure people get their value. It's a scenario where both the scale and the diversification mattered."

Jonathan is upbeat about the future of Bishop's Orchards. He believes that the diversity of their crop production and marketing practices will help the business remain successful even if weather becomes a more important risk factor as climate change intensifies through mid-century.

Jonathan Bishop has been active in local and state civic and agricultural organizations for many years. He has served as a member of the USDA Farm Service Agency State Committee and is currently on the board of New Haven Farms, a non-profit organization that promotes health and community development through urban agriculture in New Haven, CT. In 2001, Bishop's Orchards was named the Mass Mutual National Family Business of the Year. Bishop's Orchards was one of sixty American farms and ranches selected for the USDA-SARE publication *The New American Farmer: Profiles of Agricultural Innovation.*

Challenges to the Adaptive Capacity of Sustainable Tree Fruit and Nut Production

The experiences of the sustainable fruit and nut growers featured in this book seem to be strongly place-based. Some growers have experienced increased production risks related to warmer and more variable winter and spring weather, while others identify changing markets and government regulations as having more significant impacts on management decisions. Growers in the West share concerns about current or future impacts of warmer winters on summer water supplies, while those in the Midwest have experienced damaging flooding from more frequent and intense heavy rainfalls. When asked about the future, some growers were upbeat about their ability to sustain production as climate conditions change, while others expressed concern about changing weather outpacing their ability to adapt.

Endnotes

1. Tilth Producers is a member of the Washington Tilth Association, a non-profit advocacy group that works to support and promote biological sound and socially equitable agriculture. tilthproducers.org/about-us/awards/

2. R. Evans. 1999. Overtree Evaporative Cooling System Design and Operation for Apples in the Pacific Northwest. USDA-ARS-Northern Plains Agricultural Research Laboratory, Sidney, MT.
3. D. Layne. 2013. Tree Fruit: Protecting Your Investment. *American/Western Fruit Grower,* September/October.
4. R. Snyder and J. Melu-Abreu. 2005. Frost Protection: Fundamentals, Practice and Economics. UN Food and Agriculture Organization, Rome.
5. Regional landscape features like mountain ranges and large bodies of water can interact to create microclimates with ideal growing conditions, like the "frost-free" fruit growing region of Colorado or the "fruit belts" in Michigan and Central Washington. Fruit belts are prominent on the eastern shores of the Great Lakes, most notably in western Michigan. A strong "lake effect" dominates the climate in these regions because prevailing westerly winds pass over the large water bodies before encountering the eastern lake shore. Proximity to the lake tends to moderate air temperatures year-round, because lake temperatures lag behind air temperatures through the year in ways that create excellent conditions for fruit production along the eastern shores of the lakes. R. Schaetzl. no date. Fruit Production. web2.geo.msu.edu/geogmich/fruit.html

CHAPTER 7

Grains

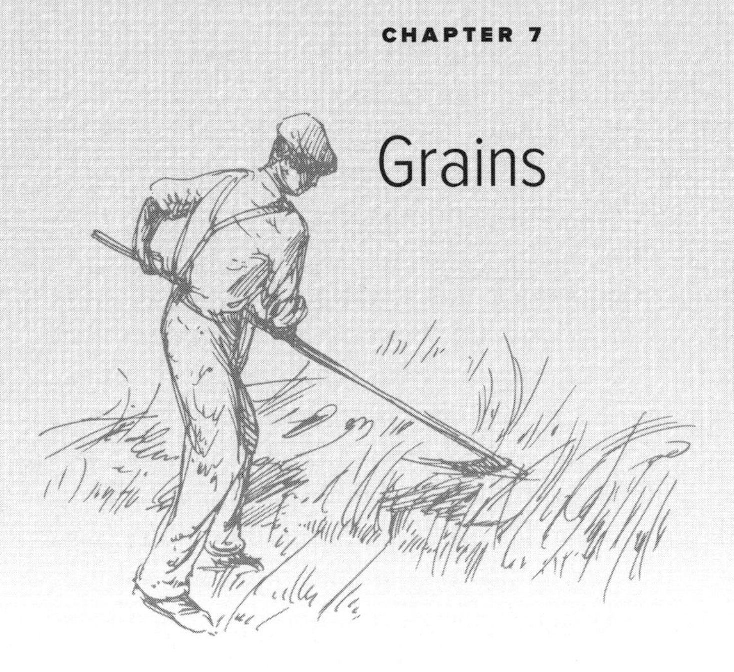

The sustainable grain producers featured in this book are located in the traditional grain-growing regions of the country. All of them managed the transition of their family farm from industrial to sustainable grain production. They represent a variety of models of successful dryland diversified grain production systems, many integrated with livestock, ranging in scale from seven hundred to five thousand acres, and they have experience managing diversified grain production at one location ranging from twenty-five to more than forty years. Some use certified organic practices, and all view soil quality, biodiversity and orientation with cooperative value-added wholesale markets as key production risk management strategies. All of these award-winning grain producers report changes in weather conditions that are outside their experience of normal weather variability and extremes.

Northwest

Russ Zenner, Zenner Family Farms, Genesee, Idaho

Russ and Kathy Zenner have been farming in the Palouse region of Idaho near the Washington–Idaho border in Genesee for more than forty years. Located about a hundred miles south of Spokane, WA, Zenner Family

Farms includes ground that was first farmed by Russ's grandfather in 1935. Kathy and Russ joined the family business in 1970 and took over management of the farm fourteen years later. In 2012, Russ's cousin Clint Zenner and his wife Alicia took on some management responsibilities, becoming the fourth generation to carry on the farming tradition of the Zenner Family in Idaho.

The Palouse is a unique region of steep rolling hills in eastern Washington and western Idaho. The rich soils and winter rains produce some of the highest dryland wheat yields in the nation. But the topography, winter rains and intensive tillage typical of past wheat production also produced some of the worst soil erosion in the nation—as much as 100 tons annual topsoil loss per acre on the steepest slopes.[1] In the 1970s, Russ became concerned about the level of erosion at Zenner Farms because of the negative impacts on soil quality and the off-farm impacts to water quality. An early innovator of reduced tillage grain production in the region, he completed a transition to a direct-seeded, no-till cropping system that greatly reduced soil erosion on the farm by 2000.

Springs are getting wetter, while the late summer, fall and winter have become drier. We had some very wet springs in '11 and '12. The spring of 2011 was by far the wettest I've ever experienced in my career. The seeding season was so wet that we were on soils when we shouldn't have been. There were a few guys who waited clear until June to plant, which is very late for our county, and the soils were still too wet.

— Russ Zenner

Today, Russ manages 2,800 acres of dryland direct-seeded crops in a three-year rotation of winter wheat, spring grains and spring broadleaf crops. Winter wheat, spring wheat, spring barley, garbanzos, lentils, peas, oilseeds and grass seed are the farm's main cash crops. Within each year of the rotation, Russ has a variety of crop types to choose from. Winter crops include soft white, club and hard red winter wheat varieties for grain or certified seed production; among the spring grain options are durum, soft white, hard white and hard red wheat varieties, malting and feed barleys and corn. Broadleaf crops include Austrian winter peas, yellow and green peas, garbanzos, lentils and oilseeds. Many considerations go into choosing a specific crop for each phase, but the overall goal is to increase the yield potential of the following crop. Over the years, the crop rotation has shifted to an emphasis on spring-seeded crops in an effort to prevent both weed and disease problems in the winter wheat crop. Other important factors in crop selection include soil type, potential markets, seasonal workloads and weather.

Russ markets about eighty percent of his crops through the Pacific Northwest Farmers' Cooperative, an important US marketer of value-added grains and dry pulses to national and international markets. Most of the remaining crops are marketed through Shepherd's Grain, a farmers' marketing alliance that sells specialty grains and flours to regional markets in the Pacific Northwest. All the growers that supply Shepherd's Grain are sustainable producers certified by the Food Alliance.

Over his lifetime on the farm, Russ has noticed some changes in weather variability and extremes. "I don't think we're seeing the temperature extremes that we saw earlier in my career, which is a benefit in a lot of regards," he says. "Twice in my lifetime we've had lows of 50 below [degrees Fahrenheit] in the Genesee Flats, once when I was just a child in the winter of 1950, and then again in the winter of 1968–69. I don't know if we've seen 20 below since then. It used to be in some of the summers, during late July and early August, we would see quite a few days of over 100 degrees. It just seems like temperatures have moderated some. This has not caused any change in our management, it's just an observation."

FIGURE 7.1. Russ and Cathy Zenner, Zenner Family Farms, Genesee, Idaho. Credit: Russ Zenner

More challenging have been changes in seasonal rainfall patterns. More rainfall in the spring and drier conditions in late summer and fall have complicated crop management at Zenner Family Farms, particularly because of the emphasis on spring-seeded crops. During the record-breaking wet spring of 2011, Russ created some soil compaction problems by planting on soils that were extremely wet, and he has been struggling to restore those soils ever since. The compacted soils may be promoting increased soil-borne disease.

Russ says that wetter springs have increased the incidence of leaf diseases in the region. "This inland Pacific Northwest region has always had pretty significant rust pressure on small grains," he explains. "Three years ago [2011] was the most significant year for rust, but to some extent, there has been a fair amount of disease pressure through the last couple of years. Whether that's weather-related or not, I don't know, but the amount of fungicide used in this region has, I would say, risen dramatically in the last few years."

Disease management has always been a challenge in direct-seeded grains in the Palouse region. As Russ transitioned the farm to direct seeding, he remembers that increased disease was a continuing issue: "In most of the things we tried, we ran into disease problems. Early on, it was because we did not have enough rotation diversity, and this still may be a big part of it. We're doing a pretty good job with our no-till system, improving organic matter, but we're not experiencing an improvement in nutrient cycling, which was one of my major goals." Russ had hoped that improved nutrient cycling would reduce production costs by reducing his reliance on purchased fertilizers.

"Longer term, no-till is not providing the results that we initially thought we might see," Russ explains. "We're still trying to identify what's holding us back. Part of it is disease. The other part of it may be the amount of manmade chemistry we're using in these systems. Many of us doing long-term no-till have wondered about this. I've questioned glyphosate, even though it is an integral tool in our no-till systems. We've got soils now that had twenty to twenty-five years of repeated glyphosate applications. There's no work being done on long-term implications of glyphosate on soil biology, and how it may possibly impact root diseases, so we just don't know. There are a lot of interactions of this manmade chemistry in the soil that we just don't understand."

Concerns about overall performance of the no-till system, plus changes in seasonal rainfall patterns, have got Russ thinking about redesigning the crop rotation to increase crop diversity and increase the proportion of fall-seeded broadleaf crops. "Years ago, we adjusted our rotations to reduce peak disease pressure in these no-till cropping systems," Russ explains. "We're planting less winter wheat than we did twenty years ago. Where it used to be 50 percent of our seeded acreage was winter wheat, we're now down to about a third. That means two-thirds of the farm is seeded in the spring, and that's a problem if it's too wet." But this plan carries some uncertainties as well, because the drier conditions in late summer and fall could complicate fall planting. According to Russ, "the late summer and early fall have been drier than during the first half of my career. That's made it somewhat challenging to get good crop establishment on fall-seeded crops."

As weather challenges increase, Russ appreciates the flexibility created by the diverse crop rotation he has developed for the farm. "Say for instance a wet spring has delayed planting, we will maybe cut back on the garbanzo acres and plant more peas or lentils instead," Russ says. "Garbanzos are the longest-maturing summer crop we have in our mix, so we can run into harvest risks in September if the crop is planted too late. Peas and lentils mature more quickly. Same way with spring grain; spring barley matures much more quickly than spring wheat. So if we need to, we can plant barley instead of wheat. And we can select from different maturity dates within the barley to fit the time available for production of the crop."

Russ is also interested to see if there are diversity benefits to the reintegration of livestock on the farm. The newest members of the management team, Clint and Alicia Zenner, have introduced a beef cattle herd that is being managed with intensive grazing of cover crops. Russ sees a lot of potential soil quality benefits associated with grazing cover crops, both from the additional crop diversity and the addition of manure to croplands. Although the cattle have only been on the farm one year, soil structure under the grazed cover crops has improved noticeably. Russ is looking forward to finding out how the grazed cover crops affect the yield and quality of subsequent grain crops.

Producing high-quality crops is an important goal at Zenner Family Farms. "I'm one of the Shepherd's Grain growers," says Russ. "If we can improve soil health, organic matter and nutrient cycling, we're going to be able to improve the nutrient density of our grains. I'm certain there will be a premium in the marketplace if we can prove our products are healthier. And to me, that would be a significant motivator to help change cropping systems management in this country." Russ is frustrated with the way that federal agricultural programs have discouraged sustainable food production. "The major portion of income transfer from the taxpayer to the producer has gone traditionally to a handful of crops. It does not encourage production diversity or reward sustainable resource management."

Russ feels strongly that government programs should encourage more nutrient output per acre. "We need to be trying to strive to improve the inherent nutritional value of the crops that come off of our ground," he says. "There is very little work, or even talk, about that. I think we lack the leadership on a national scale to recognize the incredible resource this country has had in the topsoil, in the water. That's the whole function of agriculture: to sustain humans. Why are we not more focused on maximizing the nutritional output of these cropping systems? Why aren't we using agriculture to improve human health and cut down the incredible medical costs associated with poor diet and poor nutrition, rather than relying on additives? I think there are more efficient ways to do it. I think if the taxpayer would demand their investment go in these areas, we'd be much better off."

7.1. The Nutrient Density of Food

Nutritionists have raised concerns about the declining nutrient density of the US food supply for more than thirty years.[2] The industrialization of the US food system has been accompanied by increased consumption of highly processed foods such as refined grains and other carbohydrate-rich foods with added sugars and fats. Although these foods are cheap and convenient, they are also energy-dense (high in calories) and nutrient-poor (low in nutrients). The growing consumption of these foods have been conclusively linked to rising rates of obesity and type 2 diabetes in the United States and to the global obesity epidemic. In 2005, the US Dietary Guidelines introduced the term "nutrient density" and recommended that Americans pay more attention to the nutrient density of their food choices and to select more nutrient-dense foods in every food group.[3]

Although it is clear that the kind and extent of processing plays a significant role in the nutrient density of foods, there is evidence that agricultural production conditions play a role as well. Poor-quality soils and the use of synthetic fertilizers and high-yielding modern crop varieties—all conditions associated with industrial agriculture—are all thought to play a role in the declining nutrient density of foods.[4] Although research is limited, the available data suggests general declines of between 5 and 40 percent or more in vitamins, minerals and protein in fresh, unprocessed fruits, vegetables and grains over the last fifty years.[5] A recent comprehensive review concluded that organic systems of crop production produce nutritionally superior plant-based foods,[6] and new research suggests that carefully controlled, side-by-side comparisons of crop varieties may be required to find nutritionally superior types, because no general trends in nutrient content were found when comparing heirloom and modern varieties of tomato, cabbage or broccoli grown with compost, organic or synthetic fertilizers.[7]

There is growing interest in understanding how agricultural management can play a role in reversing the observed decline in nutrient density.[8] Researchers are studying the influence of plant genetics and growing conditions on the levels of health-promoting compounds in food and working to add nutrient density information to food labels.[9] Although there is still much to be learned about how crop variety, soils, climate and farming practices interact to influence food quality, we can produce exceptionally nutrient-dense and flavorful foods in every region of the United States.[10]

Thinking about future challenges, Russ is unsure about the availability of new knowledge and technologies to effectively manage climate risk. He knows from experience that crop diversity and no-till soil management buffer yield variability from precipitation extremes. And he thinks that federally subsidized production insurance programs have been and will continue to be an important agricultural risk management tool, particularly for the transition to the next generation of farmers who have assumed high debt loads. But Russ is concerned about the lack of research and development efforts focused on agricultural adaptation. "I think the US is behind many other countries in the world. We've been to Australia several times. The first time I went there, in 2005, we visited a research station in Horsham, Victoria. There were several publications there already on climate change implications for production agriculture. I think we're trying to catch up now. I think we're behind some other countries, and that's a little bit frustrating. There just hasn't been the push to do it, I guess. I would say that would probably be one of the things that I see as maybe a limitation."

Russ has long been recognized as an innovative leader in sustainable dryland agriculture in the Pacific Northwest. A 1990 winner of the Latah Soil and Water Conservation District's Conservation Stewardship Award, Russ has been actively involved in sustainable agriculture research and education over the years through collaborations with universities and the USDA Agricultural Research Service. He is a founding member of the Pacific Northwest Direct Seed Association, which promotes the development of direct seed-cropping systems. Zenner Farms was profiled as a model US sustainable farm in the National Academy of Sciences publication, *Toward Sustainable Agriculture Systems for the 21st Century*.[11]

Great Plains

Bob Quinn, Quinn Farm and Ranch, Big Sandy, Montana

The Quinn Farm and Ranch is near the town of Big Sandy in the "Golden Triangle" of north-central Montana. The climate in this shortgrass prairie region is cold—temperatures can drop as low as forty below in winter—and dry, averaging between eight and twelve inches of rainfall, which falls mostly in May and June. Bob Quinn grew up working on his family's 2,400-acre wheat and cattle ranch established by his grandfather in 1928. After going away to college, Bob returned home with a PhD in plant biochemistry to take over management of the farm and ranch in 1978. Disenchanted with unstable commodity prices, he established Montana Flour and Grains in 1983 and begin direct marketing his wheat to bakeries. Two years later, Bob sold the cattle business so that he could focus on grain production and marketing, purchased a grain mill to add wholegrain flour to his product line and began marketing grain for his neighbors. Requests from buyers for organic grains got him interested in developing organic production methods for dryland grains, and in 1987, Bob harvested his first crop of certified organic grains. Just two years later, the entire Quinn Farm and Ranch was certified organic.

Today, Bob owns and manages 4,000 acres of certified organic land producing food grains in a full-tillage dryland production system. He manages a diverse nine-year rotation designed to build soil quality, produce crop nutrients and manage pests. The rotation includes five years of cash crops (barley, winter and spring wheat, Indian corn seed and safflower) and four years of cover crops (alfalfa, clover, peas and buckwheat). The grains and pulses produced on the farm are marketed through Montana Flour and Grains and the Kamut brand to national and international wholesale markets.

An active researcher throughout his life, Bob has worked for many years to develop new organic farming methods and crops in collaboration with Montana State University and at his own research facility on the Quinn Farm. In the 1980s, he successfully commercialized Khorasan wheat, an heirloom variety, under the Kamut trademark. Current research at the Quinn Research Farm is focused on ways to enhance the food and energy self-reliance of north-central Montana. Bob is developing a certified organic dryland cropping system that produces locally adapted vegetables and fruits as well as the fertilizer and fuel needed to

FIGURE 7.2. Bob Quinn, Quinn Farm and Ranch, Big Sandy, Montana. Credit: Kamut International

produce and process the vegetables and fruits. Other investigations currently underway include community-based biofuel production, novel oilseeds for fuel and lubricants, improving grain crop rotations and weed management, and developing salt-tolerant vegetable cultivars.

> All the time I was growing up in the decades I can remember, we have a more or less eleven-year drought cycle. We would have a drought, one or two years in eleven. The drought that started in the late nineties and went to '05 or '06 was not followed by a wet year until last year. That seven-year drought, that's extremely unusual. And then last year was extremely wet. I mean like almost double our normal rainfall. And this winter has been quite cold again, more like what we used to have, thirty-five, forty below zero. We haven't had a winter like this for twenty years probably. This is the kind of variability that I'm talking about.
>
> — Bob Quinn

Thinking about production challenges over the forty-plus years that he has been managing grain production at the Quinn Farm, Bob has seen some changes in weather over the last two decades. The growing season has lengthened, temperatures and precipitation are more variable, temperatures are warmer, and dry periods and heat waves are more common. These changes have disrupted an eleven-year drought cycle that has long been typical for the region. Bob explains, "There are more dry periods and drought, more warm temperatures and more heat waves, and there's less cold temperatures until this winter [2013–14]. In the last thirty-six years, we've had three pretty bad droughts and two pretty wet seasons, extremely wet. So we skipped one of our wet periods in the eleven-year drought cycle."

These changes, particularly the longer periods of drought and more frequent heat waves, required Bob to replace alfalfa, the main cover crop in his rotation for many years, with more drought-resistant species. "We started out using only alfalfa as a soil-building crop, and it worked out just fantastic until we entered our first major drought," Bob recalls. "We were affected by the drought a whole year earlier than all of our neighbors because alfalfa sucks so much water out of the ground. For one year, the drought actually wiped us out, so we gave up on alfalfa and went to peas as a green manure crop." Peas seemed like a good choice because they will grow enough, even in a drought year, to cover the ground and protect the soil from wind erosion and loss of water through evaporation. But after the switch, weeds began to increase and grain protein content started to drop, says Bob. "So then we went to a combination of peas and alfalfa, and then we also added clover and buckwheat. The clover is a two-year biannual crop, alfalfa's a three-year crop, and peas are a one-year crop. We added buckwheat to the rotation as a plow-down green manure crop, so now we have as much diversity in cover crops as we do in our cash crops." Bob also dropped one cash crop, lentils, from his crop rotation, because of its sensitivity to high temperatures and drought.

The performance of the new crop rotation over the last decade has reinforced Bob's appreciation for the benefits of crop diversity. "I'm very, very solid and sold on the importance of diversity as the weather has become more erratic. Some crops will do better than others, depending on when the rains come, and when the heat comes, and when the cold comes. It is all becoming more erratic, so you never know which year it's going to be. All those crops will react a little differently, some will do better, some will do worse, but if you have a big diversity, then you can save a lot of your income and your harvest overall."

Although Bob is satisfied with the performance of the mix of his crop rotation, he continues working to fine-tune the order the crops appear in the rotation to enhance yields and soil quality. "We've designed the crop rotation to have all types of different crops feeding and taking from the soil over the course of nine years," he explains. "Now we're working more on the order, trying to work out which crop best follows another crop. We're also starting to look at some companion cropping and the use of no-till on the farm. All of this ties into soil health. We believe if you have good soil health, then you'll have good plant health, and if you have good plant health, you're going to have good people health."

Thinking about the future, Bob is concerned about how more variable precipitation might affect crop production on the farm. "I'm a little bit worried about the water," he says. "The long-term climate models show us getting warmer and wetter, but I don't see the wetter happening yet for our region. Water is always a challenge because we're in a semi-arid region. That means we're always short. Normally that's the limiting factor in crop production here, so we're always looking for ways to conserve water, to catch more water that falls on the land."

Bob appreciates the increased infiltration and water-holding capacity achieved with his soil-building program, but is actively looking for ways to capture and store more rainfall, particularly during summer thunderstorms. He is also looking for different ways to conserve water because he believes that conflicts over water supplies are likely to grow in coming years. "I think that water is going to be the next big battleground. It'll make the fuel crisis look like Disneyland. If we can figure out how to grow at least some basic food crops, grains, vegetables and fruits, without a lot of water, that will be a huge benefit, because we may be forced into that at some point."

Bob also sees many potential opportunities in changing climate conditions. The Quinn Farm moved from zone 3 to zone 4 in the new USDA growing zone map released in 2012, opening up some new fruit-growing possibilities. Bob explains, "I'm pretty interested in seeing what plants can survive on the prairie and what we can do with fruit juices and that sort of thing. We are more than a thousand miles from the nearest orange tree, so why are we drinking orange juice every morning? That's what I ask my friends."

Bob has been experimenting with a new fruit juice drink that makes use of sour cherries, a fruit that can be grown locally. Sour cherry juice has all the vitamin C and antioxidants of orange juice, but is too sour to drink alone. "A mix of sweet apple cider and sour cherry juice makes an amazingly robust and zesty breakfast drink," says Bob. "People don't think about growing cherries out on the prairie, but we get a huge production from our hardy sour cherry trees. Canadian researchers have shown that if you protect fruit trees with some kind of a shelterbelt, some kind of protection against direct winds, which are pretty fierce on the prairie, you can have success with them. That's what we've done, and we are showing that to be true."

Years of active research and development work has left Bob a little frustrated at the slow erosion of support for government technical assistance programs like the Natural Resources Conservation Service and the Cooperative Extension Service. "My biggest complaint is that they do not have time to help with inventive projects. There are no provisions to help people that want to try something completely different or in a different way. They're way overloaded with these programs that someone else has designed. They don't have any time for ingenuity and something different, and I think that's too bad. They didn't use to be that way. They used to be a resource for any kind of soil conservation project that you might want to dream up."

"I wish that we could send that message to Washington somehow," Bob goes on, "that they don't have all the answers, and by pretending that they do, they really stifle imagination and response to some of these future challenges. Especially now, when things are obviously changing, we need to really be thinking of as many different solutions as possible, not trying to apply a solution that worked in the last decade. Many of those old solutions were fine, but they might not work in the future nearly as well as something that some people are only just starting to think about now, or something that hasn't even been thought of yet."

Bob has long been recognized for his innovative leadership in sustainable agriculture and organic farming research and business development. He is a recipient of AERO's Sustainable Agriculture Award (1988), was honored for a lifetime of service by the Montana Organic Association (2007) and received the National Organic Leadership Award from the Organic Trade Association in 2010. In 2011, Bob was named a Good Food Hero in recognition of his work with ancient grains, organic production and improving food quality, and in 2013, the Rodale Research Institute recognized him with a national Organic Pioneer Award. Quinn Farm and Ranch is one of sixty American farms and ranches selected for the USDA-SARE publication *The New American Farmer: Profiles of Agricultural Innovation.*[12]

Gabe Brown, Brown's Ranch, Bismarck, North Dakota

Gabe Brown has been producing cattle, feed and food grains near Bismarck, North Dakota, for more than thirty years. When he began farming on the ranch, natural resource quality was poor. The cropland had been intensively tilled for many years, soils were very low in organic matter, and light rains of as little as a half-inch an hour caused surface runoff and soil erosion. Weeds, insects, low soil moisture and poor fertility all seemed to be holding down crop yields, and the ranch's extensive native grasslands were in poor health too.

After he and his wife Shelly purchased the ranch from her parents in 1991, Gabe knew that he wanted to make some changes. He began transitioning the ranch to no-till, to diversify crop rotations and to management-intensive grazing in order to build soil quality. In 1993, Gabe converted all of his cropland to no-till, and the following year, he added peas, a legume crop, to the spring wheat, oats and barley that had been grown for many years on the ranch. Encouraged by the improvements he saw in soil quality in those first two years, Gabe planned to continue making changes to build soil quality and biodiversity on the ranch, but extreme weather caused near total crop losses at Brown's Ranch for the next four years in a row.

"Back in the mid '90s," Gabe recalls, "I went through three years of hail and one year of drought. After you lose your crops four years in a row, the banker is not going to loan you money." Short on operating funds, he didn't have much of a choice except to continue working to improve resource quality on the ranch. He didn't have the money to purchase fertilizers or pesticides for the croplands. "Since that time, I've really focused on the soil resource and on improving the water cycle, energy cycle and nutrient cycle using holistic management practices,"[13] Gabe explains. "It has been a journey, one long learning process."

FIGURE 7.3. Gabe Brown, Brown's Ranch, Bismarck, North Dakota. Credit: Brian Devore

Today, the Brown's Ranch includes about 2,000 acres of native rangeland that has never been tilled, 1,000 acres of perennial introduced forages and 2,000 acres of no-till, dryland cropland producing corn, peas (grain and forage types), spring wheat, oats, barley, sunflowers, vetch, triticale, rye and alfalfa, plus a great diversity of cover crops. Throughout the year, as many as seventy different species are planted in various fields. The grains, sunflower seeds, peas and alfalfa are sold for cash while cattle, poultry and sheep are rotationally grazed through the grasslands, cover crops and forages. No insecticides or fungicides have been used on the ranch for over a decade, herbicide use has been cut by over 75 percent, and no synthetic fertilizer has been used since 2008. Corn yields average 20 percent higher than the county average.

Water management is no longer a big issue at Brown's Ranch, where Gabe has seen first-hand the benefits of soil quality for reducing weather-related risks to production. "After no-till for twenty-plus years, very diverse crop rotations, cover crops, plus livestock integration, we've improved the health of our soil to the point that the infiltration rate, the water-holding capacity and the nutrient cycle are totally different now. Our average annual precipitation is about sixteen inches. Before, when we were only infiltrating half an inch per hour, we got very little of that water into the soil profile. We were always fighting a lack of moisture, whereas now, virtually every raindrop that falls we're able to hold." Over the twenty years that Gabe, his wife Shelly and their son Paul have worked to transition the ranch to a more sustainable production system, soil organic matter levels in the croplands have more than tripled and the soil infiltration rate has increased from one-half inch to eight inches per hour.

There is so much variability now that you really can't plan on anything. It seems that every year is an extreme. Really, I can't honestly say there is anything that we really plan on as far as the weather anymore. There is definitely more variability and precipitation from year to year and more variability in temperature.

— Gabe Brown

Gabe really noticed weather extremes getting more frequent starting somewhere around 2006 or 2007. Flooding in parts of North Dakota seems to have become the norm rather than a rare event, and more variable weather has complicated fieldwork and made crop production more difficult. "It used to be we knew we had a window of time when it's usually dry and we can harvest some forages or plant a crop," says Gabe. "We could plan for harvest during that dry period and plant crops according to a plan. That's no longer the case."

Gabe says the most effective climate-risk management tool he has is the capacity of the ranch's healthy soils to buffer more variable rainfall and temperatures. "If you can improve your soil resource and make these soils more resilient," says Gabe, "you'll be able to weather these extremes in moisture and temperature much more easily. I can easily go through a two-year drought, and it does not affect our operation to any great extent because the soil is so much more resilient. Now you're still going to have some swings in yields with annual precipitation, but it does not affect crop yields to the extent that it used to. If you have a healthy resilient soil resource and a functioning water cycle, then your crops and livestock are not nearly as susceptible to these extremes."

Gabe appreciates the flexibility his diverse crop rotation allows him in variable weather conditions. Because he plants throughout the year, he can make adjustments to fine-tune the crop rotation plan to current weather conditions. "That's the beauty of the diverse system of ours," Gabe explains. "At times, we want to plant the cover crop, and then if the weather conditions change, maybe it's dry, we'll change the mix of that species a bit for more crop types that can handle drier conditions or vice versa. We have a really big toolbox to choose from." This ability to switch out crops gives Gabe more ability to adapt his crop rotation to weather variability and extremes than other producers growing just a few crops, a point that has not been lost on the conventional producers that visit the Brown's Ranch to see for themselves how Gabe's farming system works. Gabe explains, "I tell people this when I speak in the Corn Belt. Those guys plant either corn or soybeans—that's all they plant. If corn and beans don't work out for those guys, then they're going to have a poor year, whereas we have the ability to switch in or out of so many different crops. It just makes management so much easier."

Gabe has also made changes to his livestock production to better fit it to the ranch's natural environment and to improve natural resources on the farm. He uses management-intensive grazing techniques and grazes his cattle on native prairie, improved pastures and annual cover crops. Gabe explains, "The way we manage our livestock operations, there are very few weather-related events that will affect our animals. We used to calve in February and March, so shelter, animal health and feeding during the cold were all a problem. Now, we calve in late May and June out on grass, and that is a healthy environment for them. Due to our selection process, the cattle are now more adapted to our environment. We raise cattle in a much more natural way now. The environmental extremes do not affect our livestock as much anymore."

For more than a decade, Gabe has been a popular speaker at farming conferences throughout the country. He also hosts thousands of visitors to the ranch each year and is proud to say that he has hosted visitors from all fifty states and sixteen foreign countries over the years. As weather variability and extremes have increased, Gabe has noticed a groundswell of interest in his methods from farmers and ranchers with a more conventional mindset. "The weather is always brought up at every meeting, because they are seeing more extremes and more variability and they are asking, 'How do I buffer that?' Every place I go these days I am speaking to full rooms of people because they are realizing that the conventional agriculture model just isn't working. They've been through a period here the last several years of very good commodity prices, but it's rapidly changing, and they realize they can't keep going on the way they are."

Gabe is also heartened by new connections being made between soil health and human well-being. "People are starting to see how the conventional production model is contributing to the human health crisis in this country. I think that a lot of this relates to the soil and how we've degraded it. This has led to the destruction of the water cycle, and that's severely affecting society. People are seeing they have to change. That's why soil health and regenerating our resources is so important."

Gabe expressed concern about the barriers created by crop insurance in grain production. "The current program is antagonistic to healthy soils," says Gabe. "Farmers are now making planting decisions based on crop insurance guarantees. This leads to lower diversity which negatively affects soils, and it is terrible for the consumer. Ideally, I would like to see all federal subsidies for crop insurance eliminated. If this were done, producers would quickly learn that the success of their operation depends on a healthy soil ecosystem. If we truly want to regenerate our soils, crop insurance subsidies should be eliminated." Gabe believes that programs to reward good soil management would be an effective way to encourage changes in farm management practices, but he acknowledges that such programs could create a whole lot of bureaucracy too.

Asked about the future, Gabe is confident that the resilience he has cultivated on his ranch will serve him well as climate change intensifies. He is optimistic, too, about the increased interest in his production system. "No matter what the reason," says Gabe, "weather, economics or a little of both, there is growing interest among conventionally minded grain producers in how soil health can increase production system resilience."

Gabe Brown is nationally recognized for his innovative work in soil quality and integrated production systems. He is a popular speaker at farming conferences throughout the country and regularly hosts tours and workshops at the Brown's Ranch. He has actively participated for many years in soil quality research on his ranch in collaboration with university and federal scientists. Brown's Ranch won the National Cattlemen's Beef Association's Environmental Stewardship Award in 2006, and Gabe was named the USA Zero-Till Farmer of the Year in 2007. He is the recipient of the 2008 Honor Award from the Soil and Water Conservation Society. In 2012, Gabe was honored with the Food Producer's Growing Green Award by the Natural Resources Defense Council. Gabe was a featured speaker at the 2014 National Cover Crop and Soil Health Conference, a landmark event that brought together three hundred agricultural leaders and innovators to explore how to enhance the sustainability of American agriculture through improved soil health.

Gail Fuller, Fuller Farms, Emporia, Kansas

Gail Fuller is the third generation to own Fuller Farms, located in east-central Kansas near Emporia. Gail learned about farming from his grandfathers, who were both farmers, and by working side by side with his father on their 700-acre family farm. In the late 1980s, Gail took over the grain production side of Fuller Farms. Like many producers in those times, he adopted no-till to try and reduce serious soil erosion problems and improve profitability by reducing the fuel needed for fieldwork. By the mid-1990s, the livestock had been dropped from the farm system, and Gail had expanded corn and soybean production to more than 2,000 acres by leasing neighboring land.

Thinking back on the transition to no-till, Gail recalls following best management practices of the time which involved simplifying the farming system quite a bit. "Basically corn and soybeans were the only two crops we grew. When we went to no-till, we kicked wheat and milo out of our rotation. We had a four-way rotation—corn, soybeans, wheat and milo—during the '80s, and we raised cattle, but that all got kicked out with the big rush to no-till in the '90s. When no-till first really got popular, cows and no-till weren't allowed. It was thought at the time that cattle were too destructive to soils and the damage they caused by trampling farm ground couldn't be fixed without tillage, so the cows got kicked off."

Gail says that soil erosion did not seem to get much better with the switch to no-till, perhaps because the corn–soy rotation didn't leave much crop residue. "It was all corn and soybeans," Gail explains, "and most of the corn was being chopped for silage at the time." When this is done, the entire corn plant—grain and stalk—is harvested, so there is very little plant matter left in the field. "There was zero carbon in the system," Gail recalls. "I don't have any documentation, but our erosion definitely did not get better just because we switched to no-till, because we just weren't leaving anything for the soil. And the shift to no-till created some issues with soil fertility and also increased insect and disease problems." By the late 1990s, Gail added some cover crops into the rotation and brought cattle back to the farm in an effort to build soil quality and reduce soil erosion. Although early attempts to manage cover crops within the no-till system were challenging, diverse crop rotations, integrated with cattle, have been a central feature of Fuller Farms since 2003.

FIGURE 7.4. Gail Fuller, Fuller Farms, Emporia, Kansas. Credit: Gail Fuller

Today, Gail manages a large variety of cash crops, cover crops, cattle, sheep and poultry in a highly diversified and integrated dryland production system with the goal of keeping a living root in the ground at all times. A typical cash crop mix in a given year might include winter canola, winter barley, winter triticale, winter wheat, spring wheat, corn, grain, sorghum, soybeans, sunflowers, red clover, safflower, oats and peas. Cover crops increase diversity on the farm even more, by adding thirty to forty additional species. The 75-head beef cowherd is intensively grazed on continuous cover crops, and beef cattle are finished on the farm. Gail believes that the cattle are key to his crop management and thinks of the crops and livestock on the farm as one integrated whole.

While the 80s were challenging weather-wise, they don't hold a candle to what we've had here in the last 13 or 14 years. The extremes that we are going through right now are really extreme. Obviously time erases all those memories, but I don't remember the wild swings like we've seeing today, and definitely the 90s were not like this.

— Gail Fuller

Since Gail diversified the farming system, fertility and pest challenges created by the shift to no-till are, as Gail puts it, "in the rearview mirror." Occasional crop nutrient or pest problems are easily managed these days in the well-established and highly diverse farming system. What is becoming increasingly challenging, according the Gail, is the weather. He started noticing greater extremes in temperatures and precipitation around 2000, as best as he can recall. He remembers the '90s as just a little more settled and predictable, as well as a little wetter than normal; "1993 and '95 were extremely wet years," he says and laughs. "Maybe on a grand scale we were starting to see these wild swings in the '80s and '90s. They've just become much more defined and much more sudden. Instead of having prolonged periods of below or above average, we're just going over cliffs all the time."

Flooding, dry periods and drought have all become more frequent and intense in the last ten to fifteen years, according to Gail. "For instance, we had one of our biggest floods in recorded history in November of '98, and then we've been dry since, with the exception of '08 and '09, which I think were the two wettest years in history in our area. Then we had close to the two driest years ever in '11 and '12. This last go-round of drought has just been unlike anything we've ever seen. Starting in June of 2010, we have been in a pattern of six to eight months with no precipitation and then we get it all at once. Our last big round of precipitation was in August of last year [2013], when we had 18 inches of rain in 16 days. August is normally our driest month." More extremes in temperatures have also interfered with crop production at Fuller Farms in the last few years. "As an example, August of 2013 was one of the coldest on record, and it was followed by the hottest September on record. The shifts right now are just really becoming challenging."

While these changes in precipitation have increased the complexity of managing crop and livestock production at Fuller Farms, warmer spring and summer temperatures, particularly warmer nights, have definitely reduced crop yields, according to Gail. "The winter grains like cooler nights. They mature during late May and June, and normally we're already getting pretty warm by then. We also get a lot of humidity, so it's pretty hard to cool off at night." In 2012, temperatures were so much warmer all through the spring and summer that everything was about thirty days early. The winter grains were stressed during the grain filling period by the hot, summer-like conditions in spring, while corn and soybeans were stressed throughout the summer by excessive heat. Gail recalls, "It was just over 100 degrees every day, day after day that summer. If we could have cooled off at night and let those plants relax a little bit, we probably would have had a better chance. It's really the nighttime temperatures that got us more than anything. Obviously 110 degrees in the day is not anything to like, but when they can't cool off at night, it makes it so much tougher the next day."

Other farmers in the region have noticed the more variable weather, but they have not been affected because a loophole in crop insurance allows them to "double dip", according to Gail. "In 2012, which was the hottest, driest year since 1936, producers that had corn in their rotation had one of their best financial years ever."

Although Gail takes advantage of crop insurance too, he views crop diversity as his best insurance against crop failure. He has been an innovator of extremely diverse cover crop polycultures, called cover crop cocktails, mixes of many different species that are tolerant to many environmental conditions. "For us the multispecies mixes of cover crops have been the slam dunk," Gail explains, "just so obviously building resilience into your system. When you're putting ten or twelve things in the mix—or fifteen or twenty things in the mix is even better—something's going to survive, whatever you throw at it. That keeps the system alive. It keeps the microbial community fat, the earthworms fat. It just keeps the whole system operating much healthier and much more resilient than just a monoculture crop out there."

7.2. Cover Crop Cocktails

The use of cover crops is a very common sustainable agriculture practice. Cover crops are plant species that are actively managed on the farm but are not sold into markets. Instead of contributing directly to farm income, cover crops provide many other important benefits, such as supplying crop nutrients; suppressing weeds, insects or diseases; building soil quality; protecting soil nutrients from loss; and providing forages.[14]

The use of extremely diverse mixtures of cover crops in polyculture of four or more species is capturing the attention of sustainable and industrial farmers all over the nation.[15] Farmers are leading this new innovation in grain crop production in an effort to restore the health of soils that have been degraded by years of industrial agricultural practices.[16] Cover crops and crop rotation are also being used to address the growing problem of herbicide-resistant weeds in no-till industrial farming systems.[17]

The specific mix of species varies with farm system, planting season and objective, but the aim is to maximize the capacity of the mix to flourish no matter the weather or soil conditions. High species diversity is the key to success, so cocktail mixes typically include at least one species from each of these categories: warm- and cool-season grasses, warm- and cool-season broadleaf species and legumes.

In February 2014, the first National Conference on Cover Crops and Soil Health, organized by the Sustainable Agriculture Research and Education Program (SARE), brought together three hundred leaders in farming, academia, government, agribusiness and natural resource conservation to

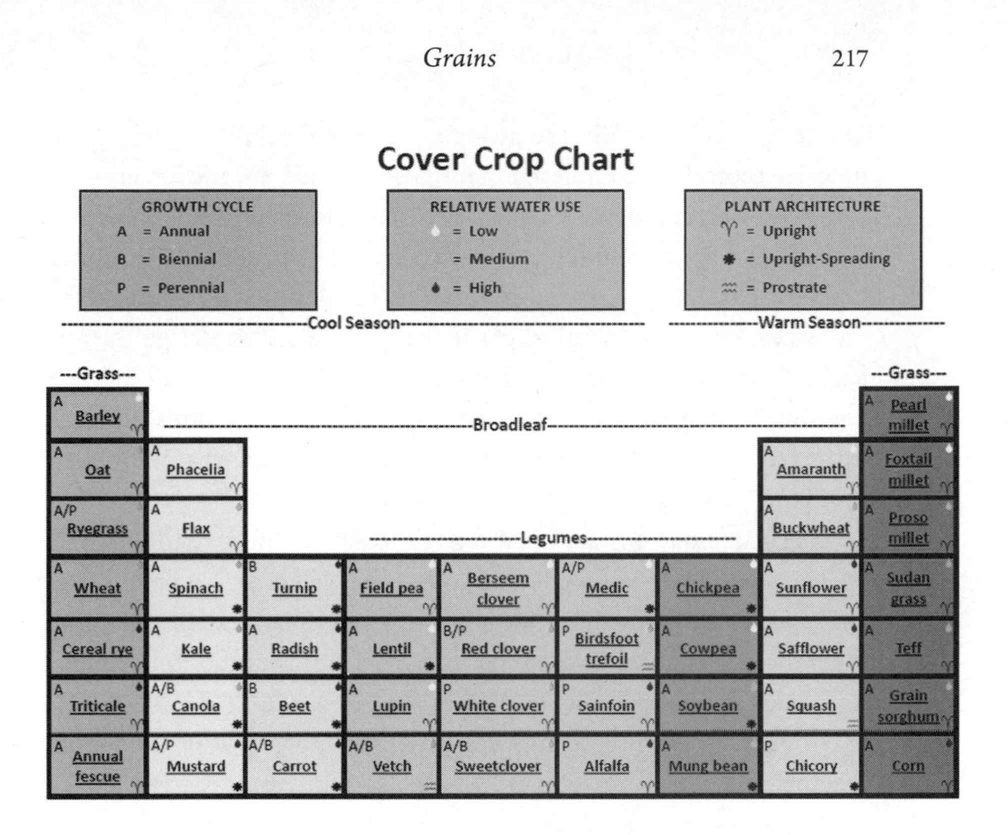

FIGURE 7.5. The Periodic Table of Cover Crops. Credit: USDA-ARS

discuss how widespread use of cover crops in industrial farming systems could contribute to a more sustainable American agriculture.[18] In conjunction with the conference, about six thousand farmers took part in Cover Crops and Soil Health Forums at Natural Resources Conservation Service and Cooperative Extension offices all across the country to view the conference presentations and participate in local conversations about cover crops and soil health. Participants in these forums shared experiences and discussed strategies—such as policy incentives and barriers, research and education—to help encourage the adoption of practices that promote soil health on farms using industrial practices.

The opening sessions and presentations from the conference can be viewed at the SARE website,[19] and a report on the local forums summarizing lessons learned and recommendations is forthcoming.

The cover crop cocktails also give Gail a lot of flexibility in crop planning. He typically plants eleven months out of the year, and having such a diverse selection of species to choose from allows him to design cover crop mixes to meet multiple goals and adjust the species to take advantage of seasonal weather conditions. "The first thing we do is look at the resource concern we have in a particular field before we design the mix for it," Gail explains. "Are you going to graze it? What time of the year are you planting it? Our knowledge of what these cover crops can do is extremely limited, but we're getting a better grasp with what we've seen here in the last four or five years with the extreme dry."

Gail goes on to explain the design process. "Then, if we think this is the weather pattern we're going to have, then we know that some cover crop species are going to handle it better than others, and so we'll make them the dominant species in the mix. And that's another thing about how a mix increases resilience. When you guess wrong, there's still going to be something there that will grow. The more diversity we put in that mix, the more it protects us." Gail feels he has just started to scratch the surface of the possibilities of cover crop mixes. Although he started with two-crop or "two-way" mixes in the late '90s, fifteen- to twenty-way mixes are the norm on the farm now. Gail imagines that the diversity of species in cover crop mixes will only continue to grow as seed suppliers offer more diversity in cover crop seeds; he thinks forty-way mixes are a real possibility in the near future.

Gail is also looking at bringing more diversity to crop production with the addition of cover crop seed and winter barley to the Fuller Farms cash crop rotation. He is experimenting with producing triticale barley, cowpeas, mung beans, buckwheat seeds. "We're going to try some different things for cover crop seed companies that bring us more diversity, that make us more resilient," Gail explains. "We don't have all of our eggs in one basket. We spread our harvest period out, so we've reduced our risk, and also we get premiums for those products, so it's kind of a win-win." Gail is also experimenting this year with no-till organic production in several fields to see if he can add additional value to his products and diversify into yet another premium market.

But resilience at Fuller Farms is not based just on crop diversity. Gail views the livestock as integral to the system as a whole. Until 2013, Gail's cattle grazed only annual cover crops, so choosing covers that are good forages is an important consideration. And the livestock provide extra crop insurance, because they can consume cash crops when weather conditions reduce quality or yield. "The livestock have now become part of the crop rotation," Gail explains. "I've got a twelve-month grazing plan. You have to have a plan if you're going to be grazing cropland. You need to have something in mind about where the cows are going to go in the worst-case scenario. In the future, the cows will be my crop insurance. That's how I plan to get income off of failed crops. I can graze them, and then we will have added value to the crops—and to the cattle. We've also recently brought sheep into the operation, because they're more drought-tolerant than cattle, so we're even diversifying there as well."

The cattle also improve the adaptive capacity of a farm or ranch by providing a value-added product and bringing all the soil quality benefits associated with management-intensive grazing, such as the stimulation of soil microorganisms and the distribution of manure back to the soil. And because Gail is grazing short-season forages throughout the year, flexible scheduling of cattle finishing allows him to capture seasonally high market prices. Gail explains, "We have value added because we are grass-finishing, but because we're grazing cover crops, I think we can expand our marketability too. In the grass-finished business, most guys are marketing in the summer. We've expanded our grazing thirty days on both ends at least, maybe sixty days in the fall, so we can finish cattle later in the year."

But there is a downside to all this diversity. In 2012, Gail landed in the center of a national crop insurance controversy. Weather conditions made it impossible for him to kill a cover crop within the time required by insurance guidelines, and his crop insurance was cancelled because his use of cover crop cocktails was interpreted as intercropping—a prohibited practice.[20] In an ironic twist, Gail had previously been awarded a Conservation Innovation grant by the USDA's Natural Resources Conservation Service (NRCS) for the same fields deemed out of compliance by the crop insurance company. After his request for a review of his case was denied, Gail took his concerns up the chain of command until he was invited to Washington, DC, to discuss the situation with USDA agency directors and his congressional representatives. Gail remembers telling the group, "We've got three government agencies controlling production agriculture, and none of them are on the same page, and that needs to stop. You need to pick a direction. We can't be getting paid here for one practice, then walk across the aisle and get denied benefits for doing the same thing."[21] USDA agency representatives agreed and created an interagency task force soon after their meeting with Gail to try and harmonize cover crop policies. In April 2014, the Task Force released a new cover crop policy to be used by all three agencies.[22] Only time will tell if the new policy will encourage the use of cover crop cocktails by commodity grain producers.

Another key barrier to managing crop diversity at Fuller Farms is the poor availability of a diverse range of cover crops and cash crops. Gail explains, "The availability of appropriate crop varieties has been a big one for us. We are switching to non-GMO varieties, and we've gone from a two-crop rotation to about twelve different crops in a five-year rotation. Just finding out what works in this area is a challenge. We're growing things like winter barley and other unusual crops for our area. So we're bringing crops from outside our region, and the varieties that are available don't always line up really well."

Although the advice and support of federal and state technical personnel and programs would be welcomed, Gail says they have not done the kind of research that he needs to improve his farming system. "We've for the most part walked away from our state university because their research is so far behind. It would be nice if we had it, if they had done the research we needed a decade ago and had this available today, that would've been phenomenal. It would've been extremely helpful, because I could have avoided all the failure that I've had here in my farm in the last ten years." Gail views the agricultural universities and many federal agricultural support programs, including crop insurance, as more of a hindrance than a help to enhancing the sustainability of US agriculture.

Thinking about the future, Gail says that the improvement he has seen on his farm and in his bottom line gives him a lot of confidence that he will enjoy continued success. And he is already planning ahead for more intense weather variability and extremes. For the longer term, he is exploring what he calls "pasture cropping," the no-till planting of an annual cash grain crop into a perennial pasture that has been knocked back by grazing or mowing. Gail sees a lot of potential for enhancing natural resource quality with pasture cropping, particularly as a long-term solution for sequestering carbon, building soil quality, enhancing the water cycle and increasing energy flow. For now, he sees a lot of adaptive potential in his farming system because he can make changes in the crop rotation to fit seasonal weather conditions. If drought conditions intensify in his region as projected, Gail thinks there will be a high demand for forages to feed cattle, because corn is not productive in hot, dry conditions. "I would probably pull out things like corn if droughts increase," says Gail, "and put in more forage-type crops that do well in those conditions. Hay and forages will become a premium, so I think we can probably do a lot of custom grazing and find ways to turn adversity into dollars."

Gail's innovative design and management have earned him both regional and national recognition. He is a regular speaker at agricultural conferences and hosts workshops and visitors at his farm. He is featured in the USDA Natural Resources Conservation Service's *Profiles in Soil Health*, a series that showcases how some of the nation's leading farmers are managing soil health to make their farms more profitable, productive and sustainable. In 2013, Gail was one of eight Kansas farmers recognized with a Climate and Energy Award by the Climate and Energy Project, a non-profit working to promote climate change mitigation in America's Heartland. He was also nationally recognized as the recipient of the 2013 Conservation Legacy Award from the American Soybean Association.

The Midwest

Ron Rosmann, Rosmann Family Farms, Harlan, Iowa

Rosmann Family Farms is comprised of 700 acres of rolling prairie lands near Harlan in west-central Iowa. With help from their children, Ron and Maria Rosmann produce certified organic feed and food grains as well as beef, pork and poultry on land that has been in the family since 1939, when Ron's father established the farm. After graduating from Iowa State University in 1973, Ron returned to manage the farm when his father retired. Ron farmed for commodity markets for about ten years, but became uncomfortable using chemical fertilizers and pesticides because of concerns about impacts to the environment and his family's health. So Ron reverted to crop rotations that were common when his father started farming, and found that the farm could be profitable and productive. Ron went on to become an early adopter in his region of more sustainable production practices such as diversified crop rotations, intensive grazing and the Swedish deep-bedding pork production system.

Today, the Rosmann Family Farms is a certified organic crop and livestock farm. The farm integrates beef cattle and swine into a diverse five-year crop rotation designed to produce the feed needed by the livestock

FIGURE 7.6. Ron and Maria Rosmann and family, Rosmann Family Farms, Harlan, Iowa. Credit: Ron Rosmann

operations, plus feed and food grains for sale. Corn, oats, soybeans, rye, barley, turnips, pasture and hay are grown on the farm. Intensive grazing practices produce grass-finished beef from a herd of eighty-five cows and calves, and a farrow-to-finish operation produces pork from thirty sows in a Swedish deep-bedding system. The organic waste produced by the farming operations, including the manure from the hogs, is composted and returned to cropland. Ron has not purchased nitrogen fertilizer since 1982, instead getting nitrogen for his crops through crop rotation and compost. Rock phosphate is applied about once every four or five years according to soil test recommendations. Beef, pork, feed and food grains are sold through wholesale and direct markets, including an on-farm retail store, managed by Maria, that is stocked with a variety of Rosmann Family Farms products as well as goods from neighboring farms.

In the last three years, we've had these huge amounts of rain in late May and June right when we're supposed to be doing fieldwork like weed control and planting cover crops into standing corn. We can't get the job done. We can't get decent weed control and so on because the window of opportunity for getting into the field is so small. So then labor becomes an issue, and you have to put in incredible long days when the opportunity arises.

— Ron Rosmann

Thinking about the effects of weather variability and extremes on farm performance over his lifetime, Ron says that there have definitely been some changes in recent years. More heavy rainfall and extreme weather, particularly extreme temperature swings in winter, more wind and a longer growing season, have created some new management challenges. "Anything involving variability in precipitation generally is a challenge," according to Ron, "but the worst for us is excess moisture because it means you can't get your work done, whether it's baling hay, cultivating, planting, harvesting, planting cover crops, you name it. Of course, that's an ongoing problem for organic farming anyway, but it's just been exacerbated by the increased variability."

Other changes in weather include more wind and more extreme storms. "We never used to have tornados," Ron explains. "I saw my first tornado when I was forty-four years old, in '94. I've seen two, three, four tornados since then. And hail. We never used to have hail very much, maybe once every seven years. Now we get it a little bit nearly every year. We're in what's come to be called a hail belt." Another change is that temperature extremes have become more extreme. "If you come from the Midwest or have lived in the Midwest," says Ron, "you know we have extremes, but the extremes have gotten worse. An example is last Monday [January 2014], it was 18 below zero temperature with a 25-mile-an-hour wind. We had a 40-below wind chill. Five days after that, it got up to 48 degrees." Looking back, Ron thinks the greater temperature extremes may have started in the early '80s, when it was common to have weeks of 50-below wind chills and huge amounts of snow. Still, the worst snow that Ron has ever experienced was in the winter of 2010. "Our cattle were out in the fields. We couldn't get out there, and when we eventually did, we had no place to feed them," Ron recalls. "It took some pretty big equipment to get through that kind of snow because it's the wide-open prairies out here."

Besides such physical challenges, Ron thinks the recent extreme swings in temperature are creating novel stress for his cattle. "What is the effect of temperature fluctuation and extreme weather changes on rates of gains?" he asks. " I think there may be more challenges for ruminants like our cattle with weather fluctuations. It used to be that the cattle could tank up on feed because you could count on the cold spell hanging around a couple weeks," Ron explains. "Now we've got 50-, 60-degree swings in a matter of days, plus you throw the wind in. That's the biggest change I've seen in my lifetime. Yes, we've always had those variations but not happening continually. You'd have longer periods in between. These extreme swings are hard."

Longer growing seasons have increased weed management difficulties on the farm too, particularly with one weed species, giant ragweed. It has been a long-term challenge in the low-lying areas of the farm, but in the last five years or so, it began moving up the hills on parts of the farm. Ron thinks wetter weather in May and June, combined with earlier spring warm-up and the longer growing season, work together to promote the spread of the weed. He explains, "We use a ridge-till system, which is a minimal tillage system. We don't do any spring tillage until we are ready to plant. We historically are not even thinking about planting corn until May 1, but now the giant ragweed are already getting big by then. They are a tough weed to take out with cultivation."

Ron says that many neighboring farmers are also increasingly challenged by changing weather conditions. Giant ragweed is even more troublesome on their farms, and more narrow windows for fieldwork are creating other challenges. Many of his neighbors, most of whom are industrial corn and soybean farmers, have abandoned best management practices in an effort to get their fieldwork done at all. Ron explains, "Now every farmer in this region, just about, gets all their nitrogen fertilizer put on in the fall because they're worried about it raining all spring. They have got to get the work done when they can because they are farming huge numbers of acres with huge equipment. But the data says between 40 and 60 percent of the nitrogen that they put on their ground is lost every year to the groundwater." Ron also says that many farmers in his region are upgrading and expanding tile drainage systems in an effort to move standing water off their fields after more frequent heavy rainfalls. This increased drainage capacity will likely wash even more nitrogen out of the farm system.

Ron is concerned about adaptations like these, which many industrial farmers are making to address current climate challenges. He fears many such farmers are focused on addressing the symptoms of climate change rather than looking for root causes such as federal support for large-scale industrial agriculture. "They keep working on genetically engineered crops and so forth, without ever looking at what caused the flooding or extreme drought in the first place. They just don't consider the big-picture issues or how the system is creating these issues." Ron goes on, "I found that out loud and clear this past fall [2013] when I was on a panel about the World Food Prize. There was a big biotech farmer on the panel who suggested to me that our farm system isn't sustainable because it can't be scaled up. I found out later that he and his brother were paid $2.3 million in farm subsidies over a thirteen-year period, which was the most of any farm in that county. And he has the audacity to say that his system is sustainable."

Thinking about his farm as a whole, Ron says, "A testimony to the resilience of our farm system is the fact that we have never lost a crop to pests in thirty years now of no pesticides. We have never even come close to having insects or disease or anything destroy our yields. We've had stable, very good yields during all that time. That's always something that people tend to look at you with, 'Huh? I find that hard to believe.' People don't believe it. Of course, I'll be the first to say we have a lot of things we could be doing better, but still, overall, the resilience shows itself in our system. Our soil quality shines here. It really does. And we work to enhance those ecosystem services, as they're called, by continually planting more trees, more shrubs, more crops for pollinators, more windbreaks and more wildlife habitat. The diversity is what will continue to play a big role for us."

In the future, Ron would like to move away from annual grain crops and incorporate more perennial nut, fruit and berry crops, both as food and as forages for livestock. He is inspired by Mark Shepard's work adapting perennial cropping systems to Minnesota prairie landscapes.[23] Ron explains, "That's the kind of thing I would ideally like to do. But right now I'm going to put an emphasis on more water-quality buffers, increasing the diversity of our headlands, getting more pollinator plants out there and planting more trees. Right now we have two and a half acres of managed woodlands, a diverse plantation that I planted with eleven different species on it."

Over the years, Ron has engaged with diverse audiences on sustainable agriculture and rural community issues, and the Rosmann family have welcomed numerous tours, visitors and research collaborators to the farm. Ron is a founding member of Practical Farmers of Iowa, a farmer-driven research and education organization, and has been active in the Organic Farming Research Foundation as a member, board member and president. In 2006, the Rosmann family received the Spencer Award for Sustainable Agriculture, which recognizes Iowa farmers who have made significant contributions toward the stability of Iowa farming. In 2012, the Catholic Rural Life Conference presented Ron and Maria with the Isidore and Maria Exemplary Award, which honors a rural couple who exemplify fidelity to a vision and vocation of rural life that combines family, stewardship and faith.

Challenges to the Adaptive Capacity of Sustainable Grain Production

The sustainable grain producers featured in this book are responding successfully to the challenges of farming in a changing climate, in large part because of the flexibility offered by the extremely diverse crop rotations that they have developed for their local environmental conditions. These producers consistently report challenges to farm management from more variable rainfall, more frequent and intense heavy rains and more intense temperature variations.

Sustainable grain producers are upbeat about their ability to sustain profitable production on their farms and ranches under current climate conditions, but identify a number of barriers that may limit their ability to adapt to further changes, including: new weed and disease management challenges; federal program barriers to increasing crop diversity in grain production; a lack of research-based information on sustainable soil and crop management practices; little or no support for local, farm-based innovation by government technical support agencies; and a general lack of information about effective adaptation planning for US agriculture.

Endnotes

1. M. Hall et al. 1999. Agriculture in the Palouse: A Portrait of Diversity. University of Idaho Bulletin 794.
2. A. Drewnowski. 2005. "The Concept of Nutritious Food: Toward a Nutrient Density Score." *American Journal of Clinical Nutrition*, 82:721–32.
3. U.S. Department of Agriculture and U.S. Department of Health and Human Services. *Dietary Guidelines for Americans, 2010*. 7th Edition, Washington, DC: U.S. Government Printing Office, December 2010. health.gov/dietary guidelines/dga2010/dietaryguidelines2010.pdf
4. B. Halweil. 2007. Still No Free Lunch: Nutrient Levels in the U.S. Food Supply Eroded by Pursuit of High Yields. The Organic Center. organic-center.org/reportfiles/YieldsReport.pdf

5. D. Davis. 2009. Declining Fruit and Vegetable Nutrient Composition: What Is the Evidence? *HortScience,* 44:15–19.
6. C. Benbrook et al. 2008. New Evidence Confirms the Nutritional Superiority of Plant-Based Organic Foods. The Organic Center.
7. Working to Curb Malnutrition from the Ground Up. ag.umass.edu/news-events/highlights/growing-nutrient-dense-vegetables
8. Food Quality. 2014. Center for Sustaining Agriculture and Natural Resources. Washington State University. csanr.wsu.edu/food-quality/
9. A. Drewnowski. 2010. Nutrient Rich Foods Index Helps Identify Healthy Affordable Foods. *American Journal of Clinical Nutrition,* 2010, 91(suppl): 1095S–1101S.
10. C. Benbrook et al. 2008. New Evidence Confirms the Nutritional Superiority of Plant-Based Organic Foods. The Organic Center.
11. Board on Agriculture. Zenner Family Farms, Illustrative Case Studies, Ch. 7 in *Toward Sustainable Agriculture Systems in the 21st Century*. 2010. National Academies Press, Washington, DC.
12. Bob Quinn, Big Sandy, Montana, Western Region, in *The New American Farmer: Profiles of Agricultural Innovation,* Second Edition. 2005. USDA Sustainable Agriculture Research and Education Program.
13. Holistic management is an adaptive management strategy that provides an ecosystem-based framework for the management of agricultural businesses. Holistic managers use some unique practices, including planning for profit, holistic decision tools, managing for healthy ecosystem services and regular monitoring of personal, business and natural resource health and well-being. Holistic Management Case Studies. Holistic Management International. holisticmanagement.org/holistic-management/case-studies/
14. Sustainable Agriculture Network. 2007. Managing Cover Crops Profitably. Handbook Series, Book 9. Sustainable Agriculture Network. Washington, DC.
15. A. McGuire. 2013. Mixing the Perfect Cover Crop Cocktail. Center for Sustainable Agriculture and Natural Resources. Washington State University. csanr.wsu.edu/cover-crop-cocktail/
16. USDA-NRCS. 2013. Soil Health Matters: Innovate to Improve Soil Health. nrcs.usda.gov/Internet/FSE_DOCUMENTS/stelprdb1083163.pdf
17. E. Unglesbee. 2014. No-Till Proponents Tout Cover Crops, Rotation in Battle with Weeds. AGFAX. agfax.com/2014/01/31/till-proponents-tout-cover-crops-rotation-battle-weeds/#sthash.MpGBR5tZ.dpuf
18. National Conference on Cover Crops and Soil Health. USDA Sustainable Agriculture Research and Education Program. sare.org/Events/National-Conference-on-Cover-Crops-and-Soil-Health

19. Ibid.
20. J. Dobberstein. 2013. USDA Task Force Clearing Up Cover Crop Rules. *No-Till Farmer*. no-tillfarmer.com/pages/Feature-Articles---USDA-Task-Force-Clearing-Up-Cover-Crop-Rules.php
21. Ibid.
22. USDA-RMA. 2014. 2014 Cover Crops Crop Insurance, Cover Crops and NRCS Cover Crop Termination Guidelines; FAQs. rma.usda.gov/help/faq/covercrops2014.html
23. M. Shepard. 2013. Restoration Agriculture: Real World Permaculture for Farmers. Acres USA.

CHAPTER 8

Livestock

The sustainable livestock producers featured in this book are located in many regions of the country. Most of these producers managed the transition of their family farm or ranch from industrial to sustainable livestock production. They offer a variety of successful models of pasture-based livestock production, some producing cattle for feedlot finishing, some finishing on farm and engaging custom processing, and a few slaughtering and doing value-added processing on-farm. These producers manage farms and ranches ranging in scale from ninety-five to one hundred and thirty thousand acres and offer experiences managing pasture-based livestock production at one location that range from twenty to almost forty years. Some use certified organic practices, and all view soil quality, biodiversity, management-intensive grazing, and orientation with value-added direct and wholesale markets as key production risk management strategies. Some of these award-winning producers report changes in weather conditions that are outside their experience of normal variability and extremes, but others do not.

The Southwest

Julia Davis Stafford, CS Ranch, Cimarron, New Mexico

The CS Ranch is located on 130,000 acres of upland shortgrass prairie at the foot of the Sangre de Cristo Mountains in northeastern New

Mexico. Cattle and quarter horses have been the focus of production since the ranch was established by Frank Springer in 1873. Today, Julia Davis Stafford and her five siblings, Springer's great-grandchildren, work together to manage cattle production and marketing, farming, hunting and quarter horse production.

Julia was raised on the ranch and has actively worked with her family to manage the cow/calf and stocker enterprises for more than thirty years. She takes the lead on strategic planning and water resource management for the ranch, and manages cattle production on the headquarters division near Cimarron. Julia uses planned grazing practices to raise cattle on native grasslands and improved hayfields, which are irrigated from the Cimarron River. For many years, the cow herd numbered between 2,500 and 3,000 head, but fifteen years of continued drought have forced Julia to destock the ranch, and today the herd is down to about 850 head. CS Ranch sells cattle mostly into wholesale markets with some direct sales locally.

Over the years, long-term weather challenges on the ranch have included variability in precipitation, dry periods and drought. Because grassland production depends entirely on precipitation, either as rain or snow, dry periods and drought are challenging because the grasslands are so responsive to variations in precipitation. Wind also creates some challenges because it tends to both dry out grassland through evaporative loss and cause soil erosion. Variability in winter snow is particularly challenging because the snowpack that builds up over winter in the mountains is the main source of river water on the ranch.

FIGURE 8.1. Julia Davis Stafford, CS Ranch, Cimarron, New Mexico. Credit: Julia Stafford

"New Mexico is very arid to begin with, and cyclical drought is very common here, so what I think of as our average annual precipitation is about fourteen to sixteen inches of rainfall," Julia explains. "That's what we hope for. Most of our ranch is upland shortgrass prairie, and we have a little bit of irrigated ground along the rivers that we mostly use to graze and raise hay for winter feed. Keeping the hayfields alive in times of drought is really tough. So that's led to us selecting varieties that are drought-tolerant and trying to minimize tillage so that we can increase soil organic matter and develop better soil health to make the most of what moisture we do get."

The hayfields used to be flood irrigated, but over the years, water-efficient, center-pivot irrigation has been installed in most of them. The water supply on the ranch is almost entirely from surface waters fed by meltwater from the winter snowpack in the nearby mountains. "The winter snowpack has been slim to none over the past ten years," said Julia. "Over the last decade of drought, the flood-irrigated areas have received water only sporadically. So a lot of the improved grass species, the bromes and orchard grass and those sorts of species, have disappeared, because we simply run out of water and can't irrigate enough to keep them alive."

We have several rivers that run through the ranch, and during all of my childhood and young adolescence the rivers were always flowing. You could count on them as a source of water for livestock. That has definitely changed over the last few years. The rivers now routinely dry up in stretches, and that has been devastating in terms of pasture use. So we have had to really scramble to address our water system where always before the rivers ran through most of the pastures.

— Julia Davis Stafford

Julia has noticed many other changes in weather in the past decade or so, particularly more variable precipitation and more extreme drought, warmer winters and more wind. "Over time we tend to go in about ten-year cycles," she explains. "But I think this drought has been longer than the last recorded cycle." Julia has also noticed that winters have gotten warmer since she was a kid. "I couldn't tell you exactly how much warmer in terms of degrees or anything, but it does seem that the winter temperatures have gotten warmer and we have less snow. Summer temperature is possibly warmer too, but that hasn't struck me as being as noticeable as the wintertime temperature changes." Winds, always a part of life in northern New Mexico, are different these days as well, according to Julia. "It seems like when I was a kid that wind blew mostly in the spring, and the month of March was always very windy, but now it seems like the real strong windy times have increased and are more common throughout the year."

These changes in weather have caused Julia to make some adjustments in production, most notably the reduction in herd size, but also in the management of the irrigated hayfields. "We've shifted very much over to a no-till type of approach under the center pivots," said Julia. "Before, when we would plow up an alfalfa field, we would plant wheat and graze it periodically before planting a hayfield again, but we are going now to less and less planting or plowing, just less soil disturbance overall. We have shifted more to no-till, and we are using perennial varieties that are good for both grazing and for making hay. The more that I've learned about soil health, the more obvious it has become that the less disturbance, the better. Having a permanent crop is better for the soil, better for the water, just better all the way around."

Julia says that other ranchers in her community perceive many of the same changes in weather. Talking about the drought is "the first and automatic topic of conversation," she says. "Everybody is bemoaning the drought. I would say that besides the drought being of tremendous concern, other ranchers also agree that we just don't have winters and the snowpack like we used to. And everyone is complaining about the wind. There is a very definite feeling of anxiety among other farmers and ranchers and townspeople around here about the lack of water, because many of the towns are facing water rationing and dwindling supplies and that sort of thing. People are leaving towns in this area and moving to metropolitan areas. I'm sure that weather is a factor in this because as agriculture decreases, business and prosperity in the area decline. There is definitely the perception that this is the worst drought that anybody has ever experienced."

Julia says that the continuing drought has created some concern about the future at the CS Ranch. "I'd say there is anxiety over wondering, 'Is this the new normal?' There is just a real awareness that if you continue to destock, at a certain point, how can the ranch keep going with fewer and fewer cattle? We are also concerned about the impact on our livelihoods and on our employees. We haven't really done any thinking ahead ten years and asking the question, What are we going to do if things keep going this way?"

Thinking about the future, Julia feels fairly confident in the management practices she uses to reduce the risks of weather variability and extremes, particularly planned grazing, soil health, water conservation and the use of drought-tolerant forage varieties and cattle that are well adapted to the region. Julia says that if climate change continues to intensify, she'll likely just continue to destock the ranch, figure out how to cut back on the need for irrigation and how to supply water to the remaining stock if surface waters were to fail.

Julia also plans to keep learning how to improve existing management practices and about new practices through participation in groups like the Quivira Coalition. "What is always tremendously encouraging to me is just the networking at these various agricultural gatherings, talking to people and going to listen to them speak," Julia explains. "Sometimes, particularly just after I get home from a Quivira Coalition conference, I feel we'll be able to sort through this and go on just fine. And sometimes I feel really anxious about how we will keep going on if these same patterns—the drops in moisture and increasing temperatures—continue. If they continue to play out on those same paths, it's going to be very tough in not very long."

Julia has been actively involved in community-based governance of regional water issues for many years. She has served on the New Mexico Interstate Stream Commission, as a board member of the Cimarron Watershed Alliance and as a member of the Western Landowners Alliance. She is an active member of the Quivira Coalition.

The Great Plains

Mark Frasier, Frasier Farms, Woodrow, Colorado

Frasier Farms is a family owned and operated ranch that spreads across 44,000 acres of rolling native shortgrass prairie in eastern Colorado, where the land is dry and the wind is almost always blowing. Brothers Mark, Joe and Chris Frasier manage the ranch in two divisions, one near Limon and the other near Woodrow. Frasier Farms produces cattle with an 800-head cow-calf and stocker operations and offer hunting leases and custom grazing. The rolling hills of native grasses such as blue grama and buffalo grass are managed using planned grazing practices that improve soil health and the health of the grassland, recycle nutrients and enhance biodiversity on the ranch.

Mark Frasier has managed the Woodrow division of Frasier Farms for more than thirty years. In addition to the calves produced on the ranch each spring and fall, Mark buys stocker cattle each spring to run about 5,000 head when fully stocked. The cattle are split into three large herds and moved through 125 pastures on the 29,000-acre ranch, leaving 90 percent of the ranch free of cattle at any one time. Cattle produced on the ranch are marketed to feedlots through value-added natural beef and source-verified programs. Mark also sells into higher-value markets by planning production to catch seasonal high prices and retaining ownership of a portion of the cattle sent to feedlots.

Mark says that year-to-year variability in seasonal weather patterns, dry periods, drought and winds are the most significant long-term challenges to dryland ranching on the eastern Colorado plains. Particularly critical is the timing and type of precipitation through the year, because grassland response to weather conditions changes throughout the growing season. Mark explains, "Let's put it this way: A two-inch

FIGURE 8.2. Mark Frasier, Frasier Farms, Woodrow, Colorado. Credit: Traci Eatherton

rain in September doesn't have near the value of two inches of rain in April or May. We just get a lot more bang from an earlier precipitation event. Spring rainfall tends to be a drizzling, all-day sort of affair in the best case. When we have the same amount of precipitation in the late summer months, it's more likely to come in an afternoon thunderstorm, so it is not as effective in terms of capturing that moisture into the soil because there is more runoff." Wind can also present some challenges to cattle management. "On the plains, especially in the spring," says Mark, "we can have some very strong winds, and we can have winds in the summer as well. If it's hot, those summer winds can just whip moisture out of the soil and have a very great drying effect on the plants and the soil."

Weather variability and extremes have always been a part of life in the Great Plains region, and Mark doesn't perceive that there has been any change in these challenges in his lifetime. "Our operation is entirely native forages," says Mark, "and the production that we get from that varies year by year per the growing condition that we experience. It has everything to do with temperature and the amount and timing of precipitation. In terms of first-day frost, in terms of wind, in terms of how long the grass stays green, all those kinds of things vary year by year. No two years are the same and they never have been. So that's really one of our challenges, trying to manage in a highly variable environment. We develop a plan, but that plan has to have significant contingencies in it because the conditions will always change."

I think historically a person would be hard pressed to say that the drought we've been in recently is any more severe than what my father experienced in the 1950s or my grandfather in the '30s. When you look directly at any one aspect of weather—variability, precipitation, temperature, length of growing season—those are always in flux. In the environment where we live, 40-degree fluctuations in temperature are not uncommon any time of the year. Our precipitation comes in concentrated periods of time, and it's not necessarily predictable. There is an inherent unpredictability about our environment.

— Mark Frasier

Mark goes on to explain that the natural environment in eastern Colorado is well adapted to the weather variability and extremes typical of the region. "For example, the grasses are extremely opportunistic," Mark says. "They don't have a narrow window within which they need to grow or to put out seed or perform some other function. They will stay in near dormancy until conditions are just right, and then they'll just explode and we'll have a significant amount of growth in a very few days. Plants have evolved in the sense that they are adapted to an environment that is unpredictable. Our environment is not ungenerous, but the plant has to be ready to grow when the conditions are right."

This same kind of preparedness to take advantage of opportunity when it comes is a central feature of Frasier Farms management. "Personally, I don't understand when people complain about not getting rain and then when it does rain oh, now it's too muddy," says Mark. "We constantly prepare for the next rain, because, generally speaking, our most limiting factor is soil moisture. If, in everything that we do, we can create an environment that is receptive to precipitation, so that whenever it does come we can take advantage of it, we will just be that much more efficient and more effective. It's an attitude or a philosophy that's grown over time. It is something a person experiences in the sense that, after you're unprepared a time or two, you begin to think ahead a bit more. So it's just a function of maturing and management, I think, as much as anything else."

Mark draws on many resources to enhance the capacity of Frasier Farms to weather variability and extremes, but the use of adaptive management strategies have proven key to his success. He explains, "I look at that key word, *variability*. You've got to have adaptive management to respond, both in terms of knowing how to respond, but also anticipating what a change will bring. Oftentimes making a timely decision is key, either in a cost-saving sense or in a sense of conservation of natural resources." Managing both a cow/calf herd and a stocker operation gives Mark the flexibility he needs to respond to changing weather conditions. "We have the two components." he explains. "The cowherd is actually a smaller piece of what we do. We're bringing in most of the livestock on the ranch. If conditions are not favorable for the growth of grass, we don't bring as many cattle to the ranch or we can destock early or in some other way change the number of cattle on the ranch. That's our control valve. We have certain performance expectations for the cattle.

If conditions are not good, then we don't meet those expectations, and on the flip side, if conditions are very good, we exceed them. It's not a doomsday situation for us to pull the cattle from the grass because they're destined for a feed yard anyway, so the fact that they may be going to the feed yard forty-five days earlier than normal or weighing fifty or a hundred pounds less than what we expected is not good, but it's not a complete disaster."

Mark also uses some other tools to reduce weather-related production risk. He finds that long-term weather forecasts—one or two months ahead—can be helpful when making stocking decisions if conditions are dry. He is also trying out a new type of federally subsidized insurance that insures the livestock producers against a large deviation from normal precipitation. This "rainfall insurance" is a pilot program of the USDA Risk Management Agency and at present is only available in a few locations.

Thinking about the future, Mark is fairly confident that he has the resources needed to keep Frasier Farms healthy, productive and profitable despite weather variability and extremes, although his confidence is related to the particular situation, as he explains. "I'd say it is really situational. It depends on what the extreme is and when and how it presents itself. There are times when I feel very compromised just because there's not much I can do at the moment. And there are other times that situations unfold more slowly, and if you have the capacity to understand what's happening, you can modify the resources you have or take advantage of opportunities to mitigate risks." He feels fortunate to have had the opportunity to learn how to successfully manage grasslands and cattle in the more variable climate of the Great Plains. "I've seen extremes in almost every sense, and so I know what comes next. I know what the end result is likely to be. I have been through it, and I am prepared to deal with it. Although no one likes to be in that position, I'd say I'm comfortable with it."

Mark is active in the civic life of his community and has provided leadership over the years to a number of community-based and agricultural organizations. He consults with other ranchers on holistic range management and is a regular speaker at agricultural events. Mark currently serves as the president of the Colorado Livestock Association. In 2003, Frasier Farms received the National Cattlemen's Beef Association Regional Environmental Stewardship Award, which recognizes the outstanding stewardship practices and conservation achievements of cattle producers across the United States. Frasier Farms was profiled as one of sixty model US sustainable farms and ranches in the USDA-SARE publication *The New American Farmer: Profiles of Agricultural Innovation.*

Gary Price, The 77 Ranch, Blooming Grove, Texas

The 77 Ranch near Blooming Grove, Texas, sits at what could be considered ground zero for climate change impact in the continental United States. Historic drought in the southern Great Plains in 2011 and 2012 led to massive destocking of beef cattle on ranches throughout the region. Declining cattle herds drove the closures of cattle feeding and processing operations in the region and cost hundreds of jobs, while rising beef prices set new records. Yet through it all, Gary and Sue Price, owners and operators of the 77 Ranch, were able to maintain production of their 190-head cowherd without the need for supplemental feed or water. What

FIGURE 8.3. Gary and Linda Price, The 77 Ranch, Blooming Grove, Texas. Credit: Karl Wolfshohl

sets the 77 Ranch apart from other ranches in the region? Why is it so resilient to drought?

The Prices began assembling their ranch almost forty years ago through the purchase of neighboring croplands and degraded rangelands. The productivity of a remnant native tallgrass prairie on the ranch in drought made a big impression on Gary. Using planned grazing, along with technical and financial support from numerous public and private organizations, Gary began to restore native prairies throughout the ranch, convinced that they could form the basis for a resilient, productive and profitable cattle production system. The historic drought of 2011 and 2012 seems to have proven him right.

Today, Gary uses planned grazing to manage a 190-cow beef herd on the restored native grasslands that dominate the 2,500-acre ranch, which also has 200 acres of cropland and about 90 acres of improved pastures. More than 40 acres of small stock ponds and small lakes provide water for livestock and waterfowl and generate extra income through the lease of fishing and hunting rights. The majority of the ranch's income comes from marketing cattle produced on the ranch into value-added wholesale cattle markets through a source-verified program.

Gary has noticed a number of changes in the weather, which used to be pretty reliable, over the years he has been raising cattle at the 77 Ranch. "We had winters, very cold winters, and then we had a good spring flush that was very predictable," he says. "We knew when our clovers were going to start growing, and we could almost predict to the day when we were going to have enough grass to stop feeding, usually about mid-March." Gary remembers that July through early September was predictably hot and dry. "You knew you needed to get your business

We should be very wet right now. We have cracks that you can put your hand in, and that's highly unusual for the middle of March. Our grass is not growing because it's been too cold and it's way too dry. We've gone now 90 days with a little less than one inch. In a time that is usually one of our wettest times. We are following 20 inches of rain which was unusual for the fall, so I mean right as we speak (March 2014) we're seeing tremendous variability. You just don't know what's around the next corner, so you have to prepare for the worst. Hope for the best of course, but you know, hope is not a plan.

— Gary Price

taken care of by middle of July," he recalls. "We always used to say, you need to just hang on until the middle of September and then you were going to get fall rains again, and then you would be okay. That was fairly predictable. October was the transition month, November was nice, and we would wean calves in October and November. We have our calving season fixed for that, but you just don't know what to expect these days. It's incredible, but you can throw all of that out now."

Gary understands that memory can be faulty, particularly when comparing changes in weather through the years. "I know there's probably some error, but with that said, I know that I've seen some things in the last five years or so that I've never seen before. I've never seen clover die in March or April for lack of water. It was much more usual to be challenged with mud in the spring and with figuring how are we going to get around and get cattle fed. That was fairly predictable." Another big change he has noticed is the intensity of dry periods and drought. "The complete loss of precipitation for periods of time is very unusual," he says. "We had a period in 2011 where we went three months with zero rain. I never saw that before. And that is not just my perception, those were records."

Gary has made a number of changes at the 77 Ranch in response to the increased variability and extremes of the last five years. He has reduced the size of his cowherd to about eighty-five percent of the maximum he knows the land can support, and he has leased additional rangeland as it becomes available near the ranch as some additional insurance against declining forage yields in times of drought. "In ranching, production challenges are always going to be related to your ability to produce forage," says Gary. "You're trying to grow as much grass as possible and do it as economically as possible. That is really the center of what we do. Producing high-quality forage takes healthy soil, and then obviously it takes water as well. So the three main considerations would be the forages, the soil and the water."

Gary is experimenting with grazing cover crop cocktails on his croplands for the first time. "We're learning to manage for less water, we're planning on less water and more heat, so we're keeping as much cover as we possibly can on the land and then trying to balance that out with our stock numbers," he explains. "These are great times in the cow business. We have the best markets ever. We want to try to take advantage of that, but not at the expense of our resource, because we'll end up hitting a brick wall. It's a short-term gain, and then you're out of grass and that will not work in the long run. We want to do any and everything we can to protect the resource. We know that's very, very important to not take these grasses down too far. We want to manage them so that they have the ability to respond to whatever rain we do get."

Although the last few years have been extremely challenging, Gary sees some opportunity in the changing climate conditions. He admits he is an optimist and says that the increased variability has helped him become a better manager. "It's given me a better understanding of the water cycle," Gary explains, "and the importance of organic matter and soil health. These are things that we sometimes took for granted in the past and did not pay enough attention to. Once you go through a deal like 2011 ... we had thirty ponds that were completely dry. We had dead fish everywhere and gates thrown open, and some pretty big lakes that were dry. The extreme drought has made it clear how important soil health is to our overall operation. It has really motivated me to learn everything I can to make the place a sponge."

Gary is pretty upbeat about the future because the ranch has such a healthy natural resource base. He also appreciates the easy access to technical assistance and support to help him maintain the productivity and profitability of the ranch under changing conditions. "We feel pretty good. We think we are on the right track. We're not wringing our hands over this. We're thankful that we chose the right track many years ago. Everything in ranching is a moving target. It's constantly changing, some things more than others. We must be very flexible in all that we do, and especially in our thinking, to adapt to changes. That's just the way it is."

Gary and Sue are both very active in civic, natural resource and agricultural organizations; regularly host visitors, field days and workshops at the ranch and collaborate in federal, state and non-profit research projects. Gary and Sue were recognized by the Sand County Foundation with the Leopold Conservation Award in 2007. In 2012, they received the Outstanding Rangeland Stewardship Award from the Texas Section of the Society for Range Management and the Southwestern Cattle Raisers Association. The National Cattlemen's Beef Association named Gary and Sue the national winners of the 2013 Environmental Stewardship Award.

The Midwest

Dick Cates, Cates Family Farm, Spring Green, Wisconsin

Dick and Kim Cates operate Cates Family Farm, a grass-fed beef farm near Spring Green in the Driftless Area of southern Wisconsin, about an hour northwest of Madison. The farm has produced pasture-based beef for more than a century and has been in the Cates family for over forty years. It includes 700 acres of managed grazing land and 200 acres of managed forest. Dick learned the business of livestock production at the Cates farm, but left home after college to gain experience ranching in Montana and managing livestock overseas in Saudi Arabia. He returned home in 1987 and took over management of the farm's beef cattle herd and, with Kim's help, made changes to improve farm profitability, including adopting rotational grazing practices, restoring a native oak savannah on their land and using intensive grazing to restore a trout stream that runs through the farm. Since 1990, the Cates have raised stocker cattle to maturity on their farm. The grazing season usually runs from early April to the end of November. The rest of the year, the cattle that are overwintered on the farm are fed hay purchased locally. The Cates direct market their pasture-raised steers as grass-finished to grocery stores, restaurants, cafeterias and households around southern Wisconsin and in the Chicago area.

> If the grass is getting shorter and it is not raining, the worst thing you can do is keep grazing. If you take the grass down too short in our part of the country, the soil will get hot and that grass will stop growing. You've got to get off it, so the ground doesn't get hot. Last summer (2012) was the worst. I had to ship cattle off to pasture, and I've never had to do that before.
>
> — Dick Cates

More variability and extremes in weather over the last ten to fifteen years have created some new management challenges at Cates Family Farm. "In my mind, we have had more high temperatures, more dry periods, more excessive rain, wet periods, so yes, more fluctuations," explains

FIGURE 8.4. Dick and Kim Cates, Cates Family Farm, Spring Green, Wisconsin.
Credit: Dick Cates

Dick. "Moisture extremes and temperature extremes, I mean big winter extremes, minus twenty, minus thirty. More snow in the winter. Back when I was a boy, we had quite a bit of snow in the winter, but then there were a lot of decades after that we didn't have very much."

Dick has noticed a change in weather variability since about 2000. "Well, there seems to be more frequent and longer dry periods as well as more floods," he says. "We have a trout stream running through the property, and occasionally it floods. Back in 2000, we had excessive rain that caused flooding. And then we had the excessive rains that caused the flooding in 2007 and 2008. Fortunately we haven't had another big flood since then, but we've had more droughts. I have kept rainfall data on my farm since 1987. I've seen more intense rainfall events and then longer dry periods. And along with those dry periods, it has been really hot. In 2004, there was record high heat here."

More snow in winter, plus rising hay prices, have sharply increased overwintering costs. Dick explains, "It is getting more difficult in the winter, because hay is more expensive than it was, and we have more snow, so we have to feed longer in the winter. I used to open the gate in

January and February, and they could find grass. That was in the '80s and '90s, but not so now. It started about four or five years ago, I can't remember exactly. We got well over a hundred inches that winter, and since then there's been enough snow in winter to keep the grass from showing."

More variable weather has also increased the complexity of managing spring grazing at Cates Family Farm. "In the spring of 2012, there was record heat, and I was in my shorts moving fence on March 15. I have a picture of it. Then in April of 2013, we got record cold. It was awful, it was still winter, it never got warm! That was the biggest challenge on our livestock. Normally, by April 1 it is starting to change, and by April 10 you have cold rain and green grass. In 2013, we couldn't graze them until the end of April, and then they were still mucking around in the mud. The cattle were very stressed during that period. The hot wears them down, but that cold weather in April when their hair is wet and they are mucking through the mud and they are not getting good forage quality, boy it was really tough on them."

Dick has made some changes to management to reduce risks from more variable weather and extremes. He has reduced his stocking rates over time to be sure that the farm can produce sufficient forage for their cattle during more frequent dry periods and droughts. He has also thought some about how to reduce hay costs, perhaps by purchasing hay from a number of different suppliers in summer or renting some land and growing his own. Another change might involve adding irrigation to some pastures that suffer more during dry periods. "I have friends who are looking at a K-Line irrigation system, the kind that you can drag around the pasture behind an ATV. That would be something that I may have to look at. Several friends have already done that."

Although Dick plans to retire in the near future, he feels confident that grass-based livestock producers in Wisconsin have the resources needed to successfully manage the increased variability and extremes in weather. Dick explains, "I've been blessed by my connections to the Wisconsin grazing networks and our Wisconsin producer groups, as well as the Farm Bureau Federation, the Wisconsin Farmer's Union and Wisconsin Grass Works. Our extension agents are really good." At the same time, Dick wonders about farmers just entering the business.

"When I was starting out," he says, "the biggest issue was marketing, and now, honestly, for new farmers I think it will be the uncertainty of the weather."

Dick and Kim Cates have been recognized for their stewardship of the natural resources at Cates Family Farm with the 1998 Wisconsin Conservation Achievement Award and the 1999 Iowa County Water Quality Leadership Award. In 2013, Dick and Kim were recognized by the Sand County Foundation with the Leopold Conservation Award.

Greg Gunthorp, Gunthorp Farms, LaGrange, Indiana

Greg Gunthorp has been raising pigs for as long as he can remember on his family farm near LaGrange, Indiana. The Gunthorp family has always raised pigs on pasture, resisting pressure to modernize when confined animal production really took off in the pork industry in the 1980s. Just after Greg and his wife Lei took over the family pork operation in 1995, pig prices hit historic lows following an especially intense period of consolidation in the industry. At that time, with pork processors paying fourteen cents a pound for live hogs, Greg found himself selling hogs for less than his grandfather had decades ago.

FIGURE 8.5. Greg Gunthorp, Gunthorp Farms, LaGrange, Indiana. Credit: Kristin Hess, Indiana Humanities, Food for Thought: An Indiana Harvest

Greg did not want to be the last in a long line of Gunthorps to grow pigs, so he began thinking about how to reach higher-value markets. Greg believed that the growing consumer interest in local foods and pasture-raised meats on both coasts would eventually spread to the Midwest. Greg knew he could raise high-quality pork on pasture, and he knew he could market it. He also knew access to processing would be a challenge, because of the concentration in the pork industry. So in 2002, Greg built a USDA-inspected processing plant on his farm, one of only a handful in the country.

Today, Greg grows and processes pasture-based pork and poultry on 225 acres of farmland managed as perennial pasture, annual forages and grain crops. Pork and poultry are outside year-round and are protected with portable huts and electric netting. The livestock are rotated through pastures, the forage and grain crops, and a small woodland. Feed grains are grown on the farm or sourced from neighboring farms, including those of his parents and a brother. The woodland and standing corn also provide some shelter and forage for the pigs in late fall and winter, and Greg encourages mulberries in the woodlands and along fence lines because of the high feed value of the fruit.

> The weather appears slightly more variable, not significantly more, but slightly more variable. I was still farming with my dad during the severe drought in '88. The drought in 2012 was worse, but I guess we were due for another one. I don't know, but the weather does appear a little bit more variable. We've always had to deal with these weather extremes. It seems like we just have to deal with them a little more often.
>
> — Greg Gunthorp

The Gunthorp Farms production system is designed to work with seasonal weather patterns. "We try not to start too early in the spring on the birds," Greg explains, "and we don't go way too late into the fall because of how difficult it becomes for us to make sure that they've got water. We try to focus production during the time of year when the pasture and forages are growing well so that the animals are on better pasture. We try to time our production to what nature does." Greg views the high soil quality on his farm as an additional plus for production as well as a buffer against more variable rainfall. "We raise a few crops, but our soils are relatively high in organic matter, even though we've got sandy soils, because we have so much pasture. Our soils are more resilient to heavy rainfalls and more erratic rainfall patterns."

Although the processing plant has been the key to the success of Gunthorp Farms, Greg admits it is a lot to manage at times. "I always tell people we really have three businesses," he says. "We have a farm, a processing plant and a meat distribution company. In order for our model to be successful, all three of them have to function relatively efficiently and work together. We slaughter and process our own pigs, chicken, ducks and turkeys. Depending on the time of year, we have eight to twelve full-time employees for our processing plant. We do slaughter, raw fabrications of chops, roasts, steaks, chicken breast and primals. We also do ground products and sausages. We have a smokehouse, and we do our bacon in there, as well as some smoked hocks, a few smoked hams and smoked sausages." Greg also does some custom slaughtering for other local livestock producers on occasion. Gunthorp Farms meats are direct marketed through an on-farm store and weekly deliveries to more than 150 high-end restaurants and meat markets in Chicago, Indianapolis and Detroit.

The processing plant has a number of energy and environmental conservation features, including a constructed wetland for wastewater treatment,[1] solar thermal preheating for the hot water used in processing, heat recovery from the refrigeration units and geothermal space heating. Solid waste from the processing plant is composted with crop residues and returned to the pastures and croplands. Greg is pleased with his efforts to recycle wastes and conserve energy in the processing plant. "We really work on it," he says. "We're doing a few interesting things. It's kind of neat actually, and it is a lot of fun. I like to play around with alternative energies."

Thinking about weather challenges, Greg says that extreme weather is pretty much a normal part of farming in his region. "Blizzards would definitely be on the list of weather challenges," he says, "along with drought, summer heat waves and very heavy rains. Excessively high winds can make it hard to keep our shelters from flying away, but blizzards top the list, because they can make it really difficult to get to the animals and make sure that they have feed, water and a dry, draft-free place to sleep. It's more the getting to them than anything, because the snow and then the drifting snow can cause us to get stuck going out there. Then it gets cold enough that you can only stay out in it for a little bit."

Although Greg thinks other farmers believe the weather has become more variable over the last decade or so, he can't say that he has noticed any significant changes in patterns over his lifetime; however, he does think the spring warm-up pattern seems to be changing. "My grandpa's rule of thumb was you didn't throw pigs out on pasture until the last week of April because you might get a little bit of snow after that, but it wasn't going to stick," explains Greg. "And that is still somewhat consistent. I remember growing up, when I was really little, my grandpa always said you 'freeze the frogs three times.' He meant that after the frogs started singing in the spring you would get a thin layer of ice on the mud puddles and the ponds three more times. And this is the thing that is getting really weird right now. In the last twelve years, one year the frogs froze twenty-one times, and another year it was like twenty-three times. Otherwise, the frogs are just about always right on. Maybe the frogs know something we don't." Although some of these changes in weather have caught Greg's attention, they have not required any changes in production practices at Gunthorp Farms.

8.1. The National Phenology Network[2]

Phenology is the study of the timing of significant biological events in the life cycle of plants and animals. These events include things like the timing of plant emergence from the soil, the return of robins in the spring or the timing of spring bloom, fruit maturity or the onset of dormancy in trees. The success of many biological interactions—such as pollination, the rearing of young and preparation for hibernation in winter—depends on the synchronous response of diverse organisms to seasonal patterns of change in temperature, light and moisture conditions. Because plants and animals are so responsive to these seasonal patterns, observing the changes in timing of significant biological events can detect very subtle climate change effects. Several phenological indicators, including songbird wintering range and the first bloom dates for lilac and honeysuckle, are included in the set of thirty leading indicators of climate change used by the US Environmental Protection Agency.[3]

FIGURE 8.6. Spring peepers (*Pseudacris cucifer*) signal the arrival of spring in the Midwest. Credit: USGS

The National Phenology Network (NPN) brings together citizens, government agencies, non-profit groups, educators and students of all ages to monitor the impacts of climate change on plants and animals in the United States by participating in the monitoring of the phenological stages of plants and animals. Nature's Notebook is an NPN program that encourages citizens to contribute to a national phenology data set by providing monitoring guidelines for more than nine hundred plant and animal species, including auditory monitoring of spring peepers.

In 2015, more than four thousand Nature's Notebook observers will take part in a nationwide effort to evaluate the accuracy of tree leaf-out predictions generated by global climate change models. The data these citizen scientists collect will be analyzed and used to help improve predictions of the onset of spring under current and future climate scenarios.[4]

Greg says that one of his biggest challenges with weather right now is longer and more intense summer and fall dry periods. He thinks this change may be connected to the increase in center-pivot irrigation in his region. "I'm 100 percent convinced that when all these guys around us turn on their center pivots, our rain becomes very, very intermittent," says Greg. "It is almost like the rainfall just goes around us. I have no data to support it whatsoever, but I'm convinced that once they turn their center pivots on, the precipitation variability increases drastically. I think the humidity from the center pivots is changing the direction of fronts and precipitation. My dad used to say it all that time, and lots of people used to think he was crazy, but there's a lot more people starting to believe it."

Thinking about the future, Greg is pretty optimistic about the continued success of Gunthorp Farms, mostly because of the high-quality natural resources in his region and on his farm. "I think we're in a part of the country that is going to be one of the last places to be severely impacted by more weather variability," he explains. "This is mostly because we have easy access to a lot of good-quality water. We don't have the issues that the Western Corn Belt has with worrying about whether they're going to end up running out of water." In addition, Greg believes that the rolling landscape on his farm and the high-quality soils created by rotational grazing and diverse cropping help to buffer the farm from extreme weather events, as does his use of standing corn and woodlands to moderate extremes in temperatures and winds.

Greg also appreciates the accumulated wisdom developed by his family over many generations of raising pigs on pasture at Gunthorp Farms. "I think we know how to deal with weather variability in animal production," Greg says. "We've always had thunderstorms. We've always had blizzards. We've always had high wind events, high rain events. We haven't had them at the frequency that we have now, but we've always had them."

Greg goes on to explain that pastured-based producers have a really different mindset compared to producers who raise animals indoors. "Pasture-based livestock producers had to build production systems that took weather into consideration from day one," he says. "The people that put up confined livestock operations were the ones that never wanted to figure out how to deal with weather challenges in the first place. It's a very different mindset when you are growing on pasture, because you're managing a system that cooperates with nature rather than trying to just build something that works regardless of whatever nature does. It's 180 degrees on the opposite end of the spectrum."

Greg Gunthorp is active in sustainable agriculture and rural social justice issues and speaks regularly at agricultural conferences, particularly on pastured-livestock production and extreme concentration in the livestock industry, and has collaborated in research on his farm. Gunthorp Farms was profiled as one of sixty model US sustainable farms and ranches in the USDA-SARE publication *The New American Farmer: Profiles of Agricultural Innovation.*

The Northeast

Jim Hayes, Sap Bush Hollow Farm, Warnerville, New York

At Sap Bush Hollow Farm, three generations of the Hayes and Hooper family produce grass-fed lamb and beef, pastured pork and poultry, all-natural wool fiber, organic honey and all-natural handcrafts in the hills of Schoharie County, New York. All these products are sold through direct markets on the farm and at local farmers' markets, and the non-perishable products are also marketed through the farm website and a regional foods website.

Jim and Adele Hayes established Sap Bush Hollow Farm in 1979 on 160 acres of upland pastures and wooded mountains near Warnerville, about an hour west of Albany, New York. With the goal of slowly building a pasture-based livestock business, they concentrated on sheep for the first decade, producing lambs for seasonal holiday markets. As they gained experience and marketing knowledge, they next expanded into pasture-based poultry, both layers and broilers, and finally added beef and pork, all in an intensive grazing management system.

Today, Jim and Adele manage a 200-ewe flock to produce their lambs and purchase all the other livestock they finish each year. The ruminants—beef and sheep—are 100 percent grass-fed and are rotated through a system of twenty paddocks on the farm, while the poultry and pigs spend their lives on pasture and are fed grain sourced from neighboring farms. The Hayes appreciate the multiple benefits of their diversified, pasture-based livestock production system to soil quality, pest management and marketing. They use no pesticides, other than worming medications when needed for the sheep, and there is virtually no soil erosion on the farm. They have not used any soil amendments, other than lime, for more than thirty years.

FIGURE 8.7. Jim and Adele Hayes and family, Sap Bush Hollow Farm, Warnerville, New York. Credit: Jim Hayes

Over the last decade or so, more variable weather and extremes have created new challenges at Sap Bush Hollow Farm. Jim and Adele have adapted to more dry periods and drought by leasing some additional pastureland to increase their capacity for forage production, and they have built ponds to provide water to every paddock on the farm. Stronger and more frequent winds are also challenging farm operations. Even though the farm is in a sheltered location, they have had to reinforce their portable poultry huts with steel bases to withstand higher winds.

In 2011, Sap Bush Hollow was right in the path of two back-to-back hurricanes—Irene and Lee—that caused catastrophic flooding in south-central New York. Jim says that the storms were an eye-opening experience for everyone in the community, particularly with respect to how quickly the road system in the area was destroyed. "The damage that those storms caused was very frightening," Jim recalls. "That really reset our thinking in a lot of ways. Within a three-hour period the stream that runs along the state road below our house flooded and gouged out the entire road, the whole fifteen feet of macadam." Jim and Adele had moved their flock of sheep to a neighboring farm that is at a higher elevation to protect them from the storm and were disturbed to learn that they could not get to them after the storm passed. Although the sheep were less than a mile away, the flooding had destroyed the two bridges between their farm and the farm where they had sheltered their flock. "With the help of neighbors, we were able to repair the bridges enough so that we could walk over them. And so we were able to go up and drive the sheep home."

> Variability in precipitation is always a challenge when you are producing livestock on pasture. It wasn't too bad here until around 2000. But since then, particularly in the last couple of years, we've seen more variability with respect to some drier periods as well as excess moisture and flooding which is causing some problems.
>
> — Jim Hayes

Loss of power after the storm was also a worry. "We didn't know how long we would be without power," recalls Jim. "I thought we would be out for weeks at least, and we have usually several tons of meat here in storage in our walk-in freezers. We have a generator that runs off our tractor, but we only have storage for about three hundred gallons of diesel fuel on the farm."

Another worry was feed. The hurricanes hit near the end of the poultry and pork production cycle, so the farm did not have a lot of feed on hand. "We were near the bottom of the line for feed," Jim recalls, "and we had maybe a thousand chickens out here and hogs which needed grain. Fortunately, the power and roads in our area were repaired quite quickly. We were lucky in that respect." Since these storms, Jim and Adele have put in a significant amount of solar power on the farm because of concerns about the reliability of the power grid. They are now looking into adding a reserve battery system to give them additional options for powering the walk-in freezer in the event of a major disruption of the electrical supply.

An indirect effect of the storms was a loss of farmers' market sales, which are a large part of their business. "My daughter goes to a farmers' market about twenty-five miles south of us, which is a little closer to New York City. That area has a lot of second homes. During Irene and Lee, they got hit pretty bad and got pretty much wiped out. That reduced our income by about thirty percent for quite a while. I would say that this year [2013] is probably the first year it's been about fully recovered."

Over the last fifteen years, heavy rainfalls have become more frequent and have increased in intensity at Sap Bush Hollow Farm. Jim and Adele have made a number of changes on the farm to try and manage the increased surface water flows during the heavy rains. "We were getting quite a bit of flow down the valley," explains Jim, "and quite a bit of groundwater coming up and saturating the areas where we keep the livestock during the winter." They built a new drainage system to redirect surface runoff and built a new barn with a raised concrete floor to provide dry shelter for livestock.

Jim and Adele have also noticed warmer temperatures, particularly in winter, and longer growing seasons, which create some new challenges and some new opportunities. "As far as normals go, we've been here a long time, and the winters are not anywhere as near as severe as they used to be," says Jim. The warmer and wetter conditions increased parasite pressures in the sheep flock. "About eight years ago or so, we started really having problems with heavy parasite loads," explains Jim. "Because of the lack of effective deworming medications, we started using the FAMACHA system which is an eyelid test that allows you to estimate the level of infestation of *Haemonchus*, which is a major parasite of sheep.

Over time, the use of the system increases the flock's natural resistance to parasites. It's a whole new system and I think it does work."

Jim also shifted to mob grazing, a special type of rotational grazing, to reduce parasite pressures. This involves managing pastures in more mature growth phases with high-intensity grazing over very short time periods. "Now we're letting the grass grow longer," Jim explains, "and we may only take 30 percent of the available forage from the top down. We have a higher residual level of thatch, and the sheep aren't grazing so close to the ground, so we're having less parasite problems." Jim has noticed some other benefits of mob grazing as well, including increased forage production, better production during dry periods, faster recovery after grazing, better weight gains and improved soil quality. "Many producers in the area are reluctant to use it because of the amount of forage that gets pounded into the dirt," says Jim, "but I think the benefits are worth it."

Jim and Adele have made some changes to capitalize on the longer growing season and to put their new barn to good use. They have shifted their lambing season from May to April. If April weather is cold and wet, they can lamb in the barn; if it is dry and warm, they can lamb in the open as they used to do in May. Earlier lambing gives the lambs more time to grow and mature during the best part of the grazing season. Jim explains, "We're looking at a longer grazing season, and we're stockpiling more for winter grazing, two changes that are really going to help us because the difference between winter grazing versus purchasing hay is almost a factor of ten, as far as cost. We hope to get a higher percentage of our animals finished before the grazing season ends, and it looks like we may be able to finish an additional batch of chickens each year as well."

Jim says that most farmers in his area have noticed similar weather changes—more variability, warmer winters, more extreme events—and have adapted in different ways to them. Grain farmers are taking advantage of the lengthening growing season by shifting to longer-season corn varieties. Vegetable farmers are putting up more high tunnels to protect their crops from more variable weather and extend the growing season. Some sheep producers are switching from wool sheep to hair sheep, which have higher resistance to parasites, as a way to manage higher parasite pressures in the longer, warmer and wetter growing seasons.

Hay producers have shifted to baleage or silage because more variable weather has made traditional hay making so difficult.

Jim and Adele's experiences over the last decade have made them realize that they are quite vulnerable to heavy rainfall and more extreme weather events. "It's come to be a major issue," says Jim. They are actively working to identify and address major farm sensitivities to more variable weather and extremes, and they appreciate the resources available to support their efforts over the last few years. "When we built the new barn we got some cost-sharing through NRCS on part of the flooring," Jim explains. Federal cost-share money also helped with the project to divert surface water flows on the farm. The Hayes have considered federally subsidized production insurance, but don't think it would be beneficial because of their product diversity and the cost. Thinking about the future, Jim laughs and says that he isn't very confident in their ability to manage changing climate conditions. "We're doing this as best as we can," he says, "but we realize that these things aren't going to go away."

Jim and Adele welcome their customers to the farm regularly and are active in civic and agricultural organizations in their region. Sap Bush Hollow Farm was profiled as one of sixty model US sustainable farms and ranches in the USDA-SARE publication *The New American Farmer: Profiles of Agricultural Innovation*.

The Southeast

Tom Trantham, Happy Cow Creamery, Pelzer, South Carolina

When Tom Trantham got into the dairy business in 1978, there were more than five hundred dairies in South Carolina; in Greenville County, where Tom's farm is located in the upstate region near Pelzer, there were thirty. Today, there are just sixty dairies in all of South Carolina, and Tom's Happy Cow Creamery is the only one left in his county. What made Tom Trantham different? Why is he still producing milk on a small family dairy farm when so many others failed?

Like many American farmers feeling the pain of consolidation in the agricultural sector the 1980s, Tom was producing a lot of milk but barely turning a profit. "I went through some really rough times in those days, we all did," he recalls. "I know there were more suicides and broken homes and divorces and bankruptcies in the '80s, because our parity was taken away from us in 1981. After that, corporate America priced our product, and whatever they said it was worth is what we got paid. You never knew what you were going to be paid or how the price was set. You didn't have any control of your product. So it went from a wonderful family life to an almost impossible life."

Although Tom had long been among the top industrial milk producers in South Carolina, rising feed and farm chemical costs and falling prices left him with few options when he was refused an operating loan in 1987. Tom could see no way to continue in the dairy business. One sunny April morning that year, his cows broke through a fence to graze a mix of ryegrass, clover and fescue that Tom had left standing because he couldn't afford the seed and fertilizer to plant a corn crop. That evening's milking yielded a two-pound increase of milk per cow, and Tom thought, 'Why not give my cows twelve Aprils a year?' After some research into annual forage crops and intensive grazing practices, he successfully guided the transition of his ninety-cow dairy from a feed-based to a pasture-based production system, dramatically lowering his costs while increasing both herd health and milk quality.

FIGURE 8.8. Tom Trantham, Happy Cow Creamery, Pelzer, South Carolina. Credit: Cooking Up a Story

The heart of Tom's "Twelve Aprils" system is the successive planting of short-lived, seasonally adapted annual crops on about 60 acres to provide his cows with high-quality forage every month of the year.[5] The forages he uses include grazing maize, sudangrass, millet, small grains, alfalfa and clover. Variables such as weather, forage needs and field-specific conditions mean that no two years are exactly alike, but on average Tom makes five to seven no-till plantings a year. Cows graze a planting once or twice, and then the forage is cut for hay or bushhogged to prepare for the following crop. Tom's Holsteins consistently top a 23,000-pound herd average, and many of them are still producing well at ten to fourteen years of age.

With the opening of the Happy Cow Creamery in 2002, Tom's transformation from commodity dairyman to specialty milk retailer was complete. Tom built the creamery in a Harvestore silo he no longer needed for storing feed. The milk travels directly from the milking parlor to the processing plant, where it is low-temperature pasteurized and whole milk is bottled. Chocolate milk and buttermilk are also made and bottled on the farm. The milk is sold into direct wholesale markets in the upstate region of South Carolina and at an on-farm store that also retails a diverse line of mostly locally sourced fresh and processed products including produce, fruits, butter, cheeses and meats.

In almost forty years of farming in upstate South Carolina, Tom can only remember one serious drought, in 1986, but he says that some drought and high temperatures are to be expected every year. The biggest change in weather that Tom has noticed is in the number and quality of summer thunderstorms. "When I started farming in '78," he says, "I remember night rains and thunderstorms in summer, and the lightning would just light up the whole sky, and we had rains, adequate rains. For the last ten or fifteen years, the thunderstorms don't seem to be the same. They are more frequent, but yet we could still have a shortage of rain during July and August. I've also seen a difference in the storms. When we first moved here, we would be out on our porch looking at these thunderstorms, and they were very beautiful. It wasn't like now, they're so harsh. I've noticed a change in the harshness of the thunderstorms, I think. I can't understand it or really put a word to it, but I know they are different."

Really, we see some drought and hot temperatures every year. This year (2013) is the first year that we haven't really had a drought. This year it has been really wet. We had the rain, but we also didn't have the sun, so we had two big problems. I'm 72 years old, and I've never seen as much rain in a year in my life, anywhere. It really affected my crops. Our hay was 9 percent protein. It would normally have been 18 or 20. Like I say, never in my life have I endured that much rain.

— Tom Trantham

Tom appreciates the flexibility the Twelve Aprils system gives him to adjust to changing weather patterns through the year. "I prepare for what I think the situation's going to be," Tom says, "and then if it doesn't work, I just bushhog it and plant something else. That's the great thing about my system." The ability to recover quickly from mistakes or the unexpected has been particularly helpful over the years. Using no-till also provides a lot of flexibility, plus it saves time and money in fuel and equipment costs. "There's always a challenge in farming," he says, "but if you make a mistake…or maybe it isn't a mistake, maybe it rained too much or it was too dry, with my system you're not set back too much. Just the number of days it takes for you to get back out there and replant. But when you've got a hundred acres of corn silage, and you lose it, you don't have another shot until next year, so you're done for. You've got to buy feed and all, and that'll break you in a heartbeat, to have to purchase feed."

Twenty-three years of diverse no-till cropping and management-intensive grazing have produced very high-quality soils throughout the farm. "The organic matter in my soil is just unreal," says Tom. "Now the way that I do that is by managing my forages so that I graze below the knee, mow below the waist (for hay) and bushhog above the waist [to control weeds and prepare the paddock for the next planting]. Now farmers think I am crazy. I just bushhogged all that feed. But this is what gives me high-quality forage, and it does great things for my soil too. When a raindrop hits my ground, it's just like a sponge. Hardpan is not a problem on my farm. When you walk on my fields, it's like you're walking on cushion."

High soil quality and diverse cropping have also maintained soil fertility and reduced pests on the farm. With the exception one year when he applied fertilizer to plots being used by researchers on his farm, Tom has not used any chemicals or fertilizers in twenty-seven years. "The one thing that I really believe in, as much as anything I'm doing, is no use of chemicals or fertilizers," Tom explains. "You can see many of my fields have less weeds than a field that's been sprayed with every kind of thing you can think of. I really like to be able to do that."

Tom is upbeat about his farm's ability to remain productive if weather variability and extremes increase as projected for his region. He views the combination of high soil quality, no-till planting, diverse short-season annuals and management-intensive grazing as a very resilient production system. "I guess it depends on the degree of weather extremes that we are talking about," says Tom, "but with my system, I am able to adjust. If one crop goes, another one's put right in. I can respond rapidly to a situation that maybe others couldn't."

But Tom remains concerned about the continued growth of industrial dairy production and the continuing decline of family dairy farms in the United States. He heard recently about a dairy farm in Indiana that is milking thirty thousand cows. "How about having thirty thousand cows in South Carolina, but in three hundred, one-hundred-cow dairies?" he asks. "Every community would benefit. I spend a lot of money in my community here. Everybody that touches an agricultural product after it leaves the farmer's hands makes money, everybody, and a lot of it."

Tom thinks back to the days when he was selling his milk to Dean Foods. "In the 1980s, the CEO of Dean Foods made a hundred-and-fifty-something million dollars in eight or ten years and then retired. That was my milk money. That's why I was bankrupt. When there is that kind of money on the top end of the product and the guy that produced it couldn't even get enough to pay his bills, that's where this country has really messed up. It's going in the wrong direction as far as agriculture. We've lost 90 percent of our dairy farms in this country since 1970. That's just not how it ought to be."

8.2. Farm Parity: A Resilient Solution?

"For many years, trends in US agriculture have been toward greater technological advances, declining margins, declining numbers of farms and increasingly larger farms. The Nation has generally encouraged and benefited from technological advances and growth in farm size, in that higher productivity has led to low and stable food prices; however, if these trends continue unabated, farm sector resiliency, rural viability, soil conservation efforts, and the world's future stable food supply may suffer."

— Government Accountability Office, 1980[6]

This remarkably prescient warning from the GAO came during the early tremors of the 1980s farm crisis. High commodity prices early in the 1970s encouraged rapid expansion in the farm sector, which led to historic increases in crop yields. By 1977, record crops combined with rising costs for fuel, seed, pesticides and other farm inputs resulted in a twenty-billion-dollar decrease in farm income, while a drop in farmland values made it difficult for farmers to secure operating loans. This challenging combination of declining credit, increased input costs and a significant drop in income left many farmers facing bankruptcy and the loss of their family farms.[7] The American Agriculture Movement formed in 1977 to organize farmer protests and to lobby the federal government for farm parity—the idea that commodity prices should support a reasonable standard of living for farm families.[8]

The concept of parity first entered into discussions of US farm policy in 1922 as a possible solution to the severe depression in the farm sector following World War I.[9] Parity can be measured in several ways but refers generally to an equitable relationship between the purchasing power of farmers and the rest of society. By the early 1930s, more than a decade of depression had begun to threaten the stability of American agriculture, and Congress finally enacted farm support programs based on price parity—the balance between the prices paid and prices received by farmers. For the next fifty years, a core goal of agricultural policy was to raise and then maintain farm parity. In 1981, the Reagan administration eliminated parity pricing in the dairy industry. Since then, the spread between farm milk price and retail price has steadily widened (see Figure 8.9). Real farm milk price, in inflation adjusted dollars, has fallen nearly every year, driving producers to produce more just to stay in business, while retail milk prices have steadily increased.[10] Although there is some evidence that the benefits of farm support programs based on price parity may have contributed to the industrialization of the US food system,[11] parity for small and mid-scale farming operations remains a rallying cry among supporters of America's family farms,[12] and may offer an effective strategy for enhancing the resilience of the US farm sector.

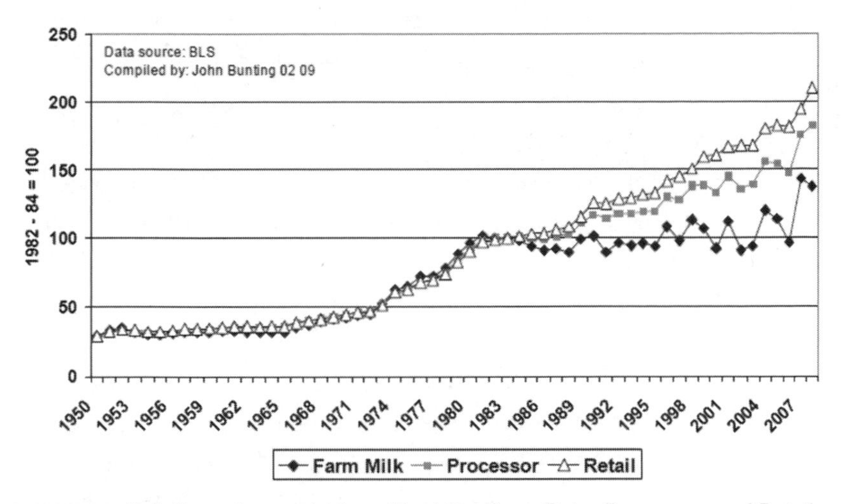

FIGURE 8.9. Relative prices paid for milk at the Farm Gate, Processor and Retail Markets, 1950 to 2009. Credit: John Bunting, National Family Farm Coalition

Today, the few remaining small-scale family dairy farms continue to struggle with rising costs of production and low milk prices. Tom understands how hard it can be to think about making radical changes when under the stress of managing an industrial dairy farm. He explains, "I know if somebody would have walked down my driveway and said, 'Hey, Tom, I want to talk to you about this Twelve Aprils dairy system,' I would have probably told them to leave. Dairy farmers just like me won't accept that because they can't take the chance. They are thinking, 'What if it doesn't work? I've got this big loan to pay and this feed bill to pay, and what if it doesn't work?' The thing about it is, it works so quick. It works so quick! It's almost an instant thing. There are so many things that are so different here that it's hard to see that it is actually easier than what they are doing now. They look at it as harder, which it just isn't. It's smooth as silk. The Good Lord woke me up back in 1987 and said, 'Tom, you've had enough. Now follow your cows. They'll show you how to be a dairy farmer.'"

Over the years, Tom has shared his experiences with countless dairy producers, researchers and policy-makers throughout the country and abroad, and he has provided leadership to many agriculture organizations. In 2002, Tom's innovative production system was recognized nationally when he received the Patrick Madden Award for Sustainable Agriculture from the USDA's Sustainable Agriculture, Research and Education Program. Tom received the 2014 Career Achievement Award from the Carolina Farm Stewardship Association. Happy Cow Creamery was profiled as one of sixty model US sustainable farms and ranches in the USDA-SARE publication *The New American Farmer: Profiles of Agricultural Innovation.*

Will Harris, White Oak Pastures, Bluffton, Georgia

Will Harris owns and operates White Oak Pastures located about ninety miles from the Gulf of Mexico in southwest Georgia near Bluffton. Established by Will's great-grandfather in 1866, Will is the fourth generation to own and manage the farm. After World War II, Will's father ran the farm using the industrial model. Will helped his father with the farm when he was growing up and then came back to the farm to take on full-time management after graduating from college in 1976. For the next fifteen years, Will continued raising calves and operating a feedlot using industrial practices, but declining profitability through the 1980s caused

FIGURE 8.10. Will Harris, White Oak Pastures, Bluffton, Georgia. Credit: White Oak Pastures

him to begin exploring higher-value alternative markets. In the mid-1990s, he began to transition the farm from industrial to grass-finished beef production using management-intensive grazing practices.

Today, White Oak Pastures produces a diversity of livestock and other products on about 2,500 acres. All the livestock produced at White Oak Pastures are pasture-raised and processed on-farm in state-of-the-art USDA-inspected beef and poultry processing plants. The plants were designed by Dr. Temple Grandin, the animal scientist renowned for her pioneering work in the humane handling of livestock. The whole farm is designed as a zero-waste system, and the inedible materials and wastewater from the processing plants are recycled back to the farmland through composting and irrigation systems. The farm is powered by a large solar array that provides about 30 percent of power needs.

With the help of his daughters Jenni and Jodi, and Jodi's husband John, Will manages the one hundred and twenty employees needed to make this diverse farming system work. Together they use multispecies rotational grazing practices to manage the production of red meats from a 700-cow beef herd, 1,100 nanny goats, a 1,000-ewe flock, 150 doe rabbits and 30 sows and poultry meats from the 320,000 chickens, turkeys, geese, ducks and guinea hens raised on the farm each year. The farm also produces organic vegetables on five acres, offers education tours of the farm and recently added on-farm dining and lodging facilities.

White Oak Pastures sells its products through wholesale and direct markets. The farm has a CSA and an on-farm store and restaurant and distributes fresh meats to grocery stores and specialty markets throughout the Southeast. All of White Oak Pasture's employees make above the minimum wage, have health insurance and get a yearly bonus.

Drought and heat have long played an important role in shaping the potential productivity at White Oak Pastures. Will thinks that temperature and precipitation have grown more variable during his lifetime at the farm, but not to such an extent that he has had to adjust any production. Part of this may be due to his long experience raising livestock and to the flexibility of raising livestock on pasture. "If I do say so myself, we're pretty good cattle people," says Will. "We've been doing it for a long, long time on the same farm. We haven't made a lot of terrible mistakes in recent years, because we learned how to do it. Our success is mostly at the mercies of the market and the weather, and this gets into some of the changes we've made." Will goes on to explain that improved soil quality on the farm has increased forage production and lowered production costs. "When I started changing the way I farm," he says, "the organic matter in my soil was less than one-half of one percent. Today it's over five percent. We're able to grow our own forage, and that takes a lot of the cost out of this compared to buying and bringing forage in."

Will uses a no-till overseeding system to produce high-quality forages for his livestock throughout the year on non-irrigated, warm-season perennial pastures. For winter grazing, Will overseeds the pastures with annual cool-season forages such as cereal rye, ryegrass and clover. "I either mow or graze very short ahead of the seeding and then broadcast annual forage seed over the pasture with a truck," he explains. "After the pasture is seeded, I come behind with a harrow that I've modified so I do not disturb the soil too much and then an aggressive drag behind the harrow to put the seed in good contact with the soil surface. Then I put the animals through the pasture to walk the seed into the ground. We don't disturb the turf much this way, and I always get a stand. I probably put a little too much seed in there, but I get a stand."

The ability to control processing on-farm and sell into high-value markets has also made the farming system more stable despite increased weather variability and extremes. "We've gone from selling live cattle to pasture-raising and on-farm processing five red meat species and five poultry species as well as organic vegetables and eggs. What we think is probably the coolest part of our story is the way the farm has come full circle in the century and a half we've had it. What we do today is so remarkably similar to what my great-grandfather and grandfather did, except we've got refrigeration, internal combustion engines and regulations. It's very similar to what they did, much more similar than what my father did and I did when I was a young man."

Will says that other farmers in his region have also noticed more variable precipitation and higher temperatures, and most have adapted by increasing their use of irrigation. "There is far more irrigation in my area than there ever used to be. Way more. The only reason I don't have it is because I'm not on a good aquifer." Will goes on to share his concern about the waste of water he has seen in his region. "Forever I've heard about water wars in the West.[13] I don't know too much about them. In the East, there's always been plenty of water for everybody, and water here is free. There's some very token permitting that's required, but there is not much to it. Once you get that, you can dig as big a hole as you want and pump as much water out as you want, and the only costs to you are the energy costs to doing the pumping and also digging the well. As a result, we grossly waste water."

Certainly drought is the most difficult for us to deal with. My father was still talking about the drought of 1954 when he died in 2000. Drought is not new here, but I do think that the variability of precipitation and temperature is increasing. I don't have a lot of irrigation, and I'm sorry I don't. We just don't and probably won't have it, because we don't have a lot of ground water here.

— Will Harris

Will has seen farmers in his area irrigate in order to moisten the soil to prepare for planting, then irrigate again to water in fertilizers and pesticides, and then water a third time after seeding the crop. "That happens a lot," says Will, "not occasionally... a lot. We're now starting to feel a little bit of competition for water with environmentalists and urban communities. I think that if things continue, and they usually do, there'll be some sort of regulation put on water use in agriculture and that will dramatically affect what we can grow. My system of farming struggles to compete with irrigated cotton and peanut farms, because they are growing subsidized crops and they gross a lot of money per acre compared to my grazing operation. It is very difficult to pay one hundred, two hundred, three hundred dollars an acre annual rent with a grazing operation, but if water becomes scarce, land rents will come down and we could afford to expand what we do."

As weather variability and extremes have increased, Will says that he has put more time and effort into looking for new opportunities that might come with the changes. "I've read a little bit about climate change," Will explains. "I don't know much about it, but it seems that for me, the bad part is more extremes in terms of temperature and rainfall, but the good news is it's generally warmer and wetter, and warmer and wetter grows stuff. If it's going to be warmer and wetter, that will benefit some of our species more than others. I have goats, sheep and cattle, and I don't know that I know what works best, but I'm going to be sensitive to it, to see what does the best. I do know that geese, guineas and ducks handle heat and cold better than the chickens. If the weather becomes warmer and there is more moisture on an annual basis, but it is more erratic, I might change the species that I grow in the pastures, both the crops as well as the livestock. I'm at least taking some notes. I'm still growing more what the market wants than what is adapting well, but I'm sensitive to that. I'll get the boot on climate change. It's either happening or it's not, but I will be watching closely to see if there are advantages we can take of it."

Will says that if the changes in weather variability increase in coming years, he will have to make some changes, particularly if dry periods and droughts intensify. "Sooner or later, I'll have to find some way to irrigate," he says. Will thinks that soil quality will continue to increase on the farm and that will help to buffer more variable rainfall. "The 5 percent organic matter of the soil will keep building," Will explains. "It already helps make the land somewhat drought-tolerant, not drought-proof, but drought-tolerant. If the variability stays about like this, we'll be okay. If it gets a little worse, we'll have to irrigate. If it's a lot worse, even irrigation won't help."

Will Harris is active in community and agricultural organizations and has been widely recognized for his innovative production system and marketing practices. He was named the Georgia Small Businessman of the Year in 2011 by the US Small Business Administration. The Georgia Conservancy named him 2012 Conservationist of the Year. Will was named the Georgia Farmer of the Year in 2012, and in 2014, Will was nationally recognized by the Natural Resources Defense Council as a recipient of the Sustainable Livestock Producer Growing Green Award.

Challenges to the Adaptive Capacity of Sustainable Livestock Production

The sustainable livestock producers featured in this book are responding successfully to the challenges of farming in a changing climate, primarily though a focus on building high soil quality and adjusting stocking levels as a strategy for maintaining production under more variable weather and extremes. Some report challenges to farm management from more variable spring weather, others from damaging storms. As a group, these pasture-based livestock producers are optimistic about their ability to sustain profitable production on their farms and ranches under current climate conditions, with the exception of the producers located along the southern tier of the country, who all share concerns about the impact of continued intense drought and heat on forage production and water supplies.

Endnotes

1. Innovations in Wastewater Management for Small Meat Processors. A Niche Meat Processor Assistance Network webinar. https://extension.org/pages/68667/innovations-in-wastewater-management-for-small-meat-processors#.U_DRuGNoHYQ
2. The National Phenology Network, usanpn.org
3. U.S. Environmental Protection Agency. 2014. Climate change indicators in the United States, Third edition. EPA 430-R-14-004. epa.gov/climatechange/indicators
4. Fact Sheet: Lifting America's Game in Climate Education, Literacy, and Training. Office of Science and Technology Policy, The White House. Washington DC. whitehouse.gov/sites/default/files/microsites/ostp/climateed-dec-3-2014.pdf
5. T. Trantham. 12 Aprils Grazing Dairy Manual. USDA. SARE. southernsare.org/Educational-Resources/Project-Products/Books-Manuals-and-Training-Guides/12-Aprils-Grazing-Dairy-Manual/Twelve-Aprils-Dairying
6. Government Accountability Office. 1980. An Assessment of Parity as a Tool for Formulating and Evaluating Agricultural Policy. A report to the Subcommittee on Family Farms, Rural Development and Special Studies, Committee on Agriculture.
7. J. Starn, 2004. Farmer Bankruptcies and Farm Exits in the United States, 1899–2002. USDA-ERS Agriculture Information Bulletin No. 788.
8. Crisis in Agriculture. nd. Nebraska Studies Website. nebraskastudies.org/1000/frameset_reset.html?www.nebraskastudies.org/1000/stories/1001_0102.htmlPlains history website
9. Government Accounting Office. 1980. An Assessment of Parity as a Tool for Formulating and Evaluating Agricultural Policy. A report to the

Subcommittee on Family Farms, Rural Development and Special Studies, Committee on Agriculture.
10. J. Bunting. 2009. The Dairy Farm Crisis: A Look Beyond Conventional Analysis. National Family Farm Coalition.

CHAPTER 9

New Times, New Tools: Managing for Resilience

The award-winning farmers and ranchers featured in the book tell of the novel production challenges associated with more variable weather, longer growing seasons, unprecedented drought, heat and flooding, and pest management that began to complicate farm and ranch management sometime in the last decade. A few have not perceived weather changes in the time they have been farming or ranching, but wonder if this is because they live in regions where weather extremes are the norm or in landscapes that serve to moderate weather extremes. All of the featured producers recognize the resilience benefits associated with sustainable agriculture practices, especially the capacity of high soil quality, biodiversity and value-added marketing to buffer the damaging effects of more variable weather and extremes.

While sustainable agriculture and agroecology offer a solid foundation of theory, principles and practices upon which to cultivate agricultural resilience, new strategies will be required to sustain agroecosystem productivity under the novel uncertainties associated with the twenty-first-century challenges of climate change and resource scarcity.[1] *Ecosystem-based adaptation*, guided by *adaptive management* oriented to *sustainability and resilience goals*, offers great potential to sustain agricultural systems in a changing world.[2]

Ecosystem-based Adaptation: Cultivating the Adaptive Power of Nature

Ecosystem-based adaptation refers to strategies that focus on conserving, restoring and using the climate protection services of ecosystems to reduce climate change vulnerability of natural and human-dominated communities.[3] An ecosystems approach to adaptation can fulfill objectives for both mitigation and adaptation to climate change, as well as build the foundation for long-term community sustainability and resilience. Ecosystem-based adaptation provides multiple benefits, both directly, through sustainable management of biological resources, and indirectly, through protection of ecosystem services.[4] For example, coastal ecosystems such as salt marshes and barrier beaches provide natural shoreline protection from storms and flooding; urban green spaces reduce the urban heat island effect and improve air quality; restored wetlands sequester carbon, contribute to a healthy regional hydrologic cycle and reduce inland flooding; and afforestation with native species facilitates the adaptation of forests to climate change.[5] Ecosystem-based adaptation has proven to be a flexible, cost-effective and broadly applicable strategy for climate change adaptation.[6] The climate protection benefits of a healthy natural resource base are widely recognized by the international development community,[7] but have received less attention in the United States.

Ecosystem management approaches are increasingly recommended in climate change adaptation efforts, for example, in protecting coasts from rising sea levels,[8] achieving global food security,[9] enhancing the climate protection services of natural landscapes[10] and fully utilizing the capacity of agriculture to provide multiple ecological, social and economic benefits to society.[11] Ecosystem-based adaptation to enhance the adaptive capacity of communities is routinely recommended as a sustainable and resilient twenty-first-century development strategy.[12]

Adaptive Management: Managing Uncertainty

A key strategy to enhance the adaptive capacity of US agriculture may be the use of adaptive management practices to support learning by doing.[13] Most commonly applied to the sustainable management of natural resources, adaptive management has more recently been applied to sustainable business and community development, including farm management. Adaptive management principles emphasize active learning about the system under management through a continuous process of evaluating and adjusting management actions based on regular observation of system behavior.[14]

Adaptive management is particularly useful in conditions of high uncertainty and complexity.[15] A key feature of adaptive management is regular monitoring of the system over time, so that management strategies can be adjusted in response to system performance. By tracking the successes and failures of different adaptive actions, managers can identify locale-specific practices that enhance system sustainability and resilience. Whole farm planning, a sustainable farm management strategy, has proven useful as an adaptive management tool in agriculture.

Whole Farm Planning: An Adaptive Management Strategy

Whole farm planning offers agricultural producers some practical tools for managing for sustainability goals in the dynamic environment of the agroecosystem. Although whole farm management was first developed to reduce agricultural impacts on environmental quality, it soon expanded to include sustainability considerations.[16] Whole farm planning makes use of the "adaptive management cycle," an iterative process involving five steps: goal setting, resource assessment, planning and implementation, monitoring progress toward goals and replanning.

Nearly all agricultural producers use performance indicators such as yield, income and profit to monitor the performance of their production system and guide business decisions; however, sustainable producers also evaluate performance on social and environmental goals.[17] Keeping sustainability goals in mind while making short-term production and marketing decisions can be challenging. Whole farm planning provides a strategy and some easy-to-use tools to aid agricultural producers managing for the multiple goals of sustainability.

Monitoring production system performance with sustainability indicators can improve farm and ranch business management in several ways. Indicators can help to clarify goals and assess options, and can be used to evaluate the success of changes in farm practices. Monitoring can be particularly useful as an early signal that a change in a management has been successful or is not going as planned. The monitoring step in whole farm planning will become even more important to sustainability goals as resource scarcity and climate change increase the uncertainty and complexity of agricultural production.

Farmers who have adopted whole farm planning feel more confident in their management decisions and report improved profitability, enhanced quality of life and increased natural resource quality on their farms and ranches.[18] Although whole farm management is a proven sustainable management tool, it is not widely used at present, even by sustainable farmers. Many farmers say they are reluctant to use these practices because of the record-keeping required for effective monitoring;[19] however, whole farm management is increasingly recommended as a sustainable management practice and is a key feature of many beginning farmer training programs.[20] The development of useful agricultural sustainability indicators has a long history of development in the United States and other industrialized nations.[21] Adaptive management strategies will likely become more important as climate risk and material resource availability create novel uncertainties to the successful management of agricultural businesses.

Understanding Resilience

Everybody seems to be talking resilience these days—but what does it really mean? Can it be measured and managed? Resilience is increasingly featured as a goal of businesses and organizations, product marketing and public policy.[22] The producers featured in this book made numerous references to resilience as they described actions taken to adapt to more weather variability and extremes and to novel pest management challenges. They also thought about their capital assets in terms of their contribution to farm resilience. Many identified soil quality, crop diversity and high-value markets as key production system features that support farm and ranch resilience to weather-related stresses, disturbances and shocks.

The concept of resilience has origins in a diverse set of disciplines, including engineering, ecology, psychology, human health and disaster management.[23] Engineering concepts of resilience focus on design for tolerance to a predetermined range of disturbance or stress. This type of resilience, called robustness, is useful in situations when the threats to the system and the system's response to those threats can be reliably predicted. The resilience concepts developed in ecology, psychology and disaster management are more relevant to agriculture because these disciplines have developed methods to assess, monitor and manage resilience as a dynamic quality of complex adaptive systems.[24]

Resilience thinking is well-grounded in ecological theory and has a long history of development in natural resource management.[25] More recently, this work has extended to the study of climate change vulnerability and adaptation in many ecological and social systems, including agriculture.[26] In the context of climate change adaptation, resilience is defined as "the ability of a social or ecological system to absorb disturbances while retaining the same basic structure and ways of functioning, the capacity for self-organization, and the capacity to adapt to stress and change."[27] The capacity to cope with, recover from and adapt to stress and change aligns with the concept of adaptive capacity as a component of vulnerability (recall that vulnerability is a function of exposure, sensitivity and adaptive capacity).[28] It is the adaptive capacity of a system that buffers the potential impact of climate change on a system. Systems with high adaptive capacity are resilient.

Key Qualities of Resilient Systems

Theoretical and practical experience has identified a number of qualities associated with resilience in natural and social-ecological systems: diversity, modularity, tightness of feedbacks and high levels of all types of capital—natural, human, social, physical and financial.[29] These qualities support the capacity for self-organization, learning, and innovation which are essential behaviors of resilient systems.

Diversity

Diversity supports the capacity for self-organization and innovation in a complex, dynamic system. Self-organization describes the spontaneous emergence of a recognizable order created by the self-interested actions of individual organisms in a system (including the people). Each independent organism in the ecosystem interacts with others to obtain desired resources, and out of these relationships, recognizable patterns begin to emerge. Self-organization made possible the characteristic patterns of the Earth's ecosystems, for example major biomes (see Figure 1.1) and early food acquisition strategies (see Chapter 1). Self-organization is also responsible for the diversity of human cultures, each with their own distinctive qualities and relationships. Sustainable producers manage natural capital assets to benefit from the self-organizing capacity of agroecosystems, for example, to produce crop nutrients (e.g., growing legumes in rotation with non-legume crops) or to suppress pests (e.g., cultivating habitat for beneficial insects).

Two forms of diversity are particularly important to resilience: functional diversity and response diversity. *Functional diversity* describes how many different kinds of species in an ecosystem are available to carry out processes such as energy flow, nutrient cycling or pest suppression. Functional diversity also describes the different kinds of professions that carry out the processes that keep community systems like food, water, energy, housing, transport and waste disposal functioning properly.

Within each functional group, there are usually a number of species (or types within a profession) that provide the same basic services, though they might do so in different ways. Sometimes these different species or types operate best in different environmental (hot, cold, wet, dry, high- or low-nutrient) conditions and may respond differently to stress or disturbance. For example, species with lower water needs can step in to keep basic ecosystem processes going during drought, while other species are available to continue these functions during a cold snap or a heat wave. An example of functional diversity in a social system is the transport options in a city: walking, bicycle, rickshaw, private car, taxi or public transport by bus or train. Depending on existing conditions, which change over time, some of these options will support transport processes better than others.

High functional diversity supports the capacity of the system to continue operating across a wide range of conditions, a quality that is called *response diversity*. Managing for high response diversity—the ability to remain productive under a wide range of conditions—is a common financial management strategy typically achieved by investing in a diversified portfolio. Sustainable producers manage for high response diversity with diversified crop rotations, integrated crop-livestock systems and diversified marketing. Management strategies that cultivate response diversity do so at the expense of efficiency, because resources are invested in components that do not directly contribute to production under all conditions. Management strategies that emphasize efficiency over response diversity do so at the expense of resilience. Another structural quality that reduces efficiency but enhances resilience is modularity.

Modularity

Modularity describes the level of connectedness of the components within a system and between the system and other systems. At a particular scale, modular systems have tight connections within each component of the whole system, but system components are only loosely connected to each other. This modular arrangement of system components increases the independence of the components and enhances the capacity of the whole system to function despite disturbances to one or more components. Modularity also encourages response diversity because each component in the system, being only loosely connected to the rest, is free to self-organize and innovate unique processes and operational strategies. A diversified sustainable farm can design each enterprise to be relatively self-sufficient and manage limited interactions between enterprises to promote the sustainability of the whole farm.

A highly connected system, in contrast, features standardized components that allow large amounts of information, energy or materials to flow freely throughout the system. This reduces the response diversity of the system, because processes are standardized, and increases its vulnerability to disturbances because they are swiftly transmitted through the whole system. An industrial farm using monoculture will experience total crop failure if any of the inputs required for production—irrigation, fertilizers, pesticides, fuel, machinery, labor—are unavailable or a damaging weather event occurs during a sensitive phase of the crop's life cycle. Large-scale, highly connected systems are vulnerable to spectacular, system-wide failures like the East Coast blackout of 2003,[30] the global financial meltdown of 2008[31] and the late blight pandemic in the Eastern United States in 2009.[32] These system-wide failures can be explained by a weakening of balancing feedbacks and a strengthening of reinforcing feedbacks as systems grow larger and more complex.[33]

Self-regulation: Balancing and Reinforcing Feedbacks

Feedbacks govern the behavior of a system through interactions between components that alter the ecological, social and economic processes in the system. These interactions involve exchanges of information, energy or materials between two or more components. Most feedbacks in ecological and social systems are balancing forces[34] that operate to maintain the system in its current state. Temperature regulation is a classic biological example of balancing feedback. If your body temperature moves outside of a normal range, temperature regulation systems send governing signals that activate responses that decrease (sweating) or increase (shivering) system temperature. Balancing forces can also occur in social systems, for example, when citizens send signals to politicians with their votes, and in physical systems, for example, a home heating system.

Much rarer in systems are self-reinforcing feedbacks, which act to push the system out of balance and to reinforce a change in the system.[35] The longer the reinforcing feedback is active, the greater its capacity to create fundamental change. Reinforcing feedbacks are powerful forces for change. These are the forces that drive growth, explosion, erosion and collapse in systems.[36] Population growth is the classic example of a biological self-reinforcing feedback: the more babies born, the more people grow up to have babies. Disease outbreaks are another: the more tomato plants infected with late blight, the more tomato plants there are to spread the disease to other tomato plants. Soil erosion is also self-reinforcing: as erosion reduces soil quality, the erosion processes increase in intensity leading to a downward spiral of soil degradation.[37] Self-reinforcing feedbacks also occur in social systems, for example the interest earned on a savings account, the erosion of the middle class or the tendency for the wealthy to get wealthier.[38] Left unchecked by balancing forces, self-reinforcing feedbacks ultimately transform the existing system into a new one with different structural and functional properties.

A complex adaptive system typically has numerous balancing feedbacks acting to maintain its integrity as conditions change. Some of the balancing feedbacks are inactive most of the time—such as the sweating and shivering feedbacks that maintain your body temperature—but are crucial to the system's well-being. As social-ecological systems develop, there is often a trend toward weakening balancing feedbacks by eliminating them (in the quest for efficiency) or by increasing the distance the feedback must travel as energy and material use and governance become more complex and increases in scale.[39] Management strategies that cultivate balancing feedbacks do so at the expense of efficiency, because resources must be invested in components that are inactive much of the time.[40] Systems that emphasize efficiency over balancing feedbacks and protect self-reinforcing feedbacks—especially those that result in the inequitable accumulation of capital assets—face a growing risk of catastrophic failure in a changing climate.[41]

Capital Assets

System resilience is associated with the amount, diversity and quality of capital assets—natural, human, social, financial and physical—to cope with, adapt to or transform a system in response to stress, disturbance and shock.[42] An abundance of high-quality capital across the full range of types offers the greatest opportunity for sustaining the system under variable and changing conditions. A healthy natural resource base provides climate protection services and enhances the response diversity of the system. High-quality human and social resources enhance the learning and innovation capacity required for innovative responses to challenging conditions. Financial and physical resources provide access to the necessary tools, equipment and technologies to put innovative solutions into action.

Holding an abundance of capital assets in reserve to enhance recovery from disturbance and shock is another feature of resilient systems. Recovery reserves can be natural (feed and forage, soil quality, biodiversity), human (experience/ease of managing change), social (community memory, knowledge, skills, assistance), financial (savings, access to financial resources) or physical (distributed energy generation, multipurpose infrastructures). Resilient systems tend to accumulate reserves

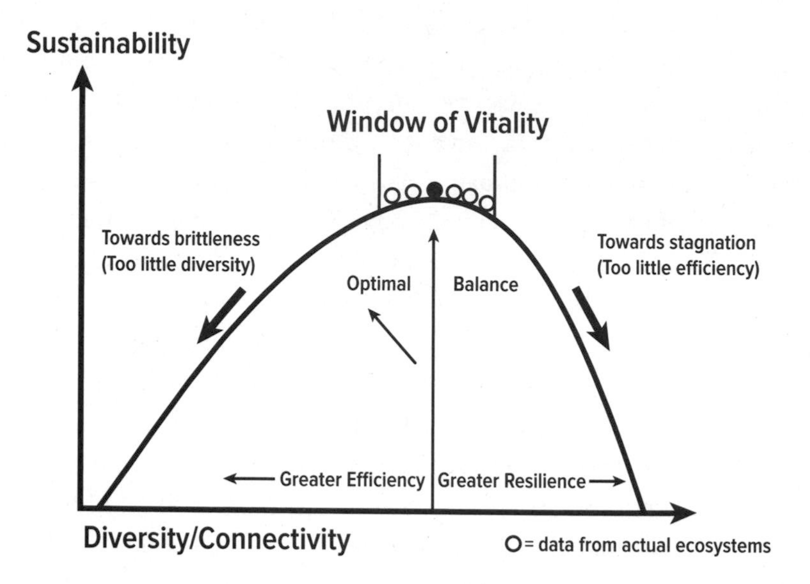

FIGURE 9.1. Efficiency and Resilience. Does the pursuit of efficiency degrade resilience? New research to develop a sustainability indicator for complex systems provides empirical evidence that it does. High diversity and connectedness provide multiple options for response should one connection be lost as a result of a disturbance; however, systems also require sufficient throughput of materials to assure crucial processes can continue. It turns out that system sustainability is highest when there is a balance of efficiency and resilience that that slightly favors resilience. Natural ecosystems tend to stay in this "window of vitality."

across all types of capital. Like managing for diversity, modularity and balancing feedbacks, investing in recovery reserves challenges traditional notions of economic efficiency because it redirects resources away from production goals; however, recent research suggests that resilience and efficiency goals are complementary and a balance of both is required to enhance the sustainability of the system[43] (see Figure 9.1).

Some Additional Resilience Considerations[44]

Scale, Identity and Desirability

Because resilience emphasizes system behavior, defining the scale and identity of the system under management is an important step in assessing and managing for resilience. The appropriate focal *scale* (scale of focus) for cultivating resilience on the farm or ranch will typically be determined by the area of land under direct management of the producer; however, resilience thinking encourages always keeping in mind the scales immediately above and below the focal scale, because processes operating at these scales (e.g., soil biology below or federal regulations above) often have a direct influence on the behavior of the whole farm or ranch system.

Identity refers to characteristic structure, function and purpose of a system. A wheelbarrow is a simple engineered system made up of some standard components (wheels, handles and a bin) organized to function (provide leverage, roll over rough ground) for a specific purpose (to move materials by hand over short distances). Unlike engineered systems, the identity of a complex adaptive system emerges as a result of self-organization and is maintained through regulating feedbacks.

Agroecosystems can be grouped into characteristic types that share many structural and functional properties, just like ecosystems (recall biomes). For example, pasture-finished beef and cattle feeding operations are very different in structure, function and purpose. Most pasture-finished beef operations will be similar in structure (different aged cattle, forages), function (energy flows from the Sun to forages to cattle and is harvested as meat) and purpose (sustainable production of beef for specialized markets). Most confined animal feeding operations also share similar structure (one narrow age range, imported grain, large fenced areas of exposed soil), function (fossil/solar energy, water, nutrients imported as cattle and feed, exported as cattle and wastes) and purpose (land- and labor-efficient production of beef for commodity markets).

Resilience practice can be used to maintain an existing system, but it can also be used to guide a transformation of the system to a new identity with a different structure, function and purpose.[45] Resilience practice brings up important questions about the *desirability* of the system. Is the system in a desirable state? Does it meet management goals and objectives? Stepping back and examining the system brings up important questions about its structure and function. Are there other ways to structure the system to improve performance? Can functions be managed differently to reduce costs or increase benefits? Resilience practice also encourages specific consideration of current or future events that might represent critical threats to the system's sustainability. If the existing system is performing well, the goal of management will be to enhance its adaptive capacity; if it is not fulfilling its purpose, resilience practice can be used to transform the system into a new, more desirable identity. Resilience practice, like all systems thinking strategies, encourages us to take a step back from the system, see it with new eyes and question underlying assumptions and rationalizations.[46]

Understanding Change: The Adaptive Cycle

Ecological theory suggests that all complex adaptive systems move through a standard cycle of development, concentration, release and reorganization. At the beginning of the development phase, conditions of high resource availability and a multitude of weak connections between components encourage rapid growth. As development continues, resource availability falls, growth slows, and fewer but stronger connections form between system components. As the resources continue to accumulate and become concentrated into a small number of strongly connected components, the system begins to lose response capacity and eventually breaks down under stress or shock. This breakdown releases resources and components to self-organize, and the cycle begins anew. Ecologists studying succession in natural ecosystems first observed the adaptive cycle, but the concept has since been observed in many kinds of complex, self-organizing systems including cultural systems and financial systems.

Adaptive cycles occur at all scales, so each component within a whole farm or ranch system moves through the adaptive cycle at its own pace, as does the system as a whole and each component of the system at the scales above the individual farm or ranch. These cycles form connections or relationships across scales that influence the behavior of components at scales above and below the focal scale. Keeping in mind the potential influence of these cross-scale relationships can be helpful when managing for resilience. For example, a system in the concentration phase of the adaptive cycle will resist innovation and change, while a system in the release and reorganization phase will welcome it.

Managing for Resilience

Resilience of two kinds can be managed in complex adaptive systems. *Specified resilience* is the resilience of some specific component of the system to a specific disturbance, for example, the resilience of the spring bloom to a late frost, the resilience of a bare soil to erosion from heavy rains or the resilience of pastures to drought in summer. *General resilience* is the capacity of a system to absorb disturbances of all kinds, including those that are novel and unexpected, so that the system maintains its structure, function and purpose.

Specified Resilience

Managing specified resilience requires answering the question: Resilience of what, to what? Specified resilience is a measure of the amount of disturbance a component of the system can absorb and adapt to without undergoing transformation to a different identity. An assessment of specified resilience involves quantifying the resilience of the system to change in three dimensions—social, economic and ecological—and across at least three scales—the focal scale and the scales immediately below and above it—in order to estimate the amount of disturbance required to push the system over a threshold and into a new identity. In theory, it is possible to directly measure the components and conditions that define specified resilience and to determine the cost/benefit of investments that enhance specified resilience. For example, the costs of constructing and maintaining a levee to protect a farm from losses due to flooding can be accurately estimated, as can the likely losses from floods (although the risks of such events are increasingly difficult to project based on historical data).

Specified resilience aligns well with the concept of sensitivity as a component of climate change vulnerability, because the sensitivities of biological and physical components of the farm or ranch system often have very specific ranges of tolerance. Most of the work to develop specified resilience management strategies has focused on natural resource management at regional and national scales.[47] While there are no specified resilience management tools suitable for use by agricultural producers, managing for specified resilience with traditional agricultural risk management strategies has proven useful; however, increasing the resilience of a system to one specific threat can reduce the resilience of the system to other threats. After identifying and managing for specified resilience, management attention to general resilience will also be required under the conditions of uncertainty associated with climate change and resource scarcity.

General Resilience

General resilience is what most people think of when they refer to resilient people, agriculture or landscapes. General resilience is the capacity of a system to absorb disturbances of all kinds, including novel or unforeseen ones, in ways that maintain the system's identity. Rather than focusing on a specific type of disturbance, general resilience describes the general coping capacity of the system. It is well-aligned with the concept of adaptive capacity as defined in vulnerability assessment.[48]

Managing for general resilience involves enhancing three key system behaviors: 1. response capacity—ability to respond quickly and effectively to buffer disturbances; 2. recovery capacity—ability to quickly restore the system after damage; and 3. transformation capacity—the ability to transition the system to a new identity when necessary.[49] It is not possible to determine the cost/benefit tradeoffs associated with general resilience; however, it is possible to learn to effectively enhance general resilience through the use of adaptive management strategies.

Cultivating a Resilient Agriculture

Agricultural sustainability and general resilience share many characteristics,[50] and taking action to enhance these qualities present many of the same challenges to agricultural managers. They are both qualities that arise through interactions between the components of the agroecosystem (including the people) and its operating context.[51] System qualities of this kind, called emergent properties, can be difficult to define or measure and are even more difficult, or impossible, to manage directly; however, sustainable whole farm management using resilience indicators offers some promising approaches.[52]

Agroecosystem Resilience Criteria

Recently, a set of research-based criteria has been proposed for the assessment of general resilience in agricultural ecosystems.[53] The proposed criteria build upon and extend sustainability principles of environmental, social and economic well-being to include a more explicit focus on the ecological and social response capacity of the agroecosystem.

Ecological Response Capacity

The six ecological criteria invite attention to the potential benefits of managing natural resource assets to enhance diversity, modularity and balancing feedbacks in the agroecosystem. These ecological criteria—self-regulation, functional and response diversity, spatial and temporal diversity, appropriate connectedness, exposed to disturbance and local natural capital—are well aligned with the ecological principles and practices of sustainable agriculture.

Ecological self-regulation describes the capacity of the ecological components of the agroecosystem to organize and create relationships that generate ecosystem services[54] with minimal management intervention. Self-regulation is promoted by an abundance and diversity of natural resource assets and the ability of plants and animals in the agroecosystem to freely engage in different relationships in their own self-interest; in other words, to self-organize. Self-regulating ecosystem processes provide balancing feedbacks to maintain the identity of the agroecosystem despite changes in internal and external conditions.

Self-regulation is enhanced in agroecosystems designed to promote high-functional biodiversity. Biodiverse agroecosystem designs typically utilize a mix of annual and perennial crops, native plants and livestock adapted to local ecological conditions to create a diverse, year-round living plant canopy that can build soil quality, provide habitat for beneficial organisms and produce a profit. A greater degree of ecological self-regulation can reduce the amount of external inputs—such as nutrients, irrigation, pesticides and energy—required to maintain agroecosystem performance.

All of the farmers and ranchers featured in this book offer models of biodiverse agroecosystems, and a few provide specific examples of

TABLE 9.1. Resilience Design Criteria for Agroecosystems. The thirteen resilience criteria listed below can be used to assess the general resilience of an agroecosystem. Sustainable agriculture practices that can enhance each indicator are listed below, along with associated sustainability indicators that can be used to assess performance of the agroecosystem on sustainability goals.

Resilience Design Criteria	Associated Sustainable Agriculture Practices	Sustainability Indicators
Ecologically self-regulated	Farm maintains diverse annual plant cover and incorporates perennials, provides habitat for beneficial organisms and aligns production with local ecological conditions	Soil Quality, Balanced Nutrient and Carbon Budget, Energy and Water Efficiency, Pest Pressure
Functional and response diversity	Diverse crop rotations, integrated and pasture-based livestock production systems, composting, alternative energy production, water harvesting	Soil Quality, Balanced Nutrient and Carbon Budget, Energy and Water Efficiency, Pest Pressure
Spatial and temporal diversity	Farm landscape is a mosaic pattern of managed and unmanaged land, diverse plant types and livestock are cultivated across space and time, diverse crop rotations integrated with livestock	Biodiversity
Appropriately connected	Collaborating with multiple producers, suppliers, markets and farmers; farm design that encourages response diversity	No comparable indicators
Exposed to disturbance	Management that accepts some controlled disturbance from weather variability, nutrient variability and pests in order to discover robust crops, livestock and production system configurations	No comparable indicators
Coupled with local natural capital	Farm builds soil quality to maintain healthy water and mineral cycles, has little need to import water or nutrients, or export waste	Soil Quality, Balanced Nutrient and Carbon Budget, Energy and Water Efficiency
Socially self-organized	Farmers and consumers are able to organize into grassroots networks and institutions such as co-ops, farmers' markets, community sustainability associations, community gardens and advisory networks	Participation and Cooperation in Community

TABLE 9.1. (cont'd.) Resilience Design Criteria for Agroecosystems.

Resilience Design Criteria	Associated Sustainable Agriculture Practices	Sustainability Indicators
Builds human capital	Investment in infrastructure and institutions to support community-based education, research and development, and local businesses, support for social events in farming communities	Time for Family Activities, Family Education, Farm Succession Plan, Local Sales, On-farm Jobs, Local Purchases, Community Cooperation, Local Identity
Reflective and shared learning	Extension and advisory services for farmers; collaboration between universities, research centers and farmers; cooperation and knowledge sharing between farmers; record-keeping; baseline knowledge about the state of the agroecosystem	Cooperation with Other Farmers, Community On-farm, Family Education
Honors legacy	Maintenance of heirloom seeds and engagement of elders, incorporation of traditional cultivation techniques with modern knowledge	Local Identity
Globally autonomous and locally interdependent	Less reliance on commodity markets and reduced external inputs; more sales to local markets, reliance on local resources; existence of farmer co-ops, close relationships between producer and consumer, and shared resources such as equipment	Local Sales, Local Purchases, Community Cooperation, Community On-farm
Reasonably profitable	The people involved in agriculture earn a living wage, a reasonable return on invested capital, have the resources needed for healthcare, education, family activities and retirement; farm work brings a feeling of satisfaction to the people working on the farm	Total Family Income, Time for Family Activities, Family Health, Satisfaction from Farming, Farm Succession Plan

the extraordinary capacity of biodiverse designs to reduce the need for purchased inputs and buffer more variable weather and extremes while accumulating natural assets. Brendon Rockey, Gabe Brown, Gail Fuller, Ron Rosmann, Jim Koan and Tom Trantham innovated vegetable, fruit, grain and livestock production systems that have little or no need for fertilizer or pesticide inputs or supplemental irrigation (except for Brendon, who increased profits and reduced irrigation needs by 50 percent), while enhancing natural resource quality on their farms.

Functional diversity and response diversity describe the capacity of the agroecosystem to maintain healthy function of the four farming system processes (energy, water, mineral, community dynamics[55]) and other ecosystem services. *Functional diversity* describes the number of different species or assemblages of species that participate in agroecosystem processes to produce ecosystem services. *Response diversity* describes the diversity of responses to changing conditions among the group of species or species assemblages that contribute to the same ecosystem function. Agroecosystems designed with high functional and response diversity have the capacity to produce ecosystem services over a wide range of environmental conditions.

Cultivating response diversity buffers the agroecosystem from ecological disturbances. Sometimes called *redundancy*, response diversity emerges through the duplication of system components and relationships that are critical to agroecosystem function. Response diversity may decrease agroecosystem efficiency, but it enhances response and recovery capacity. The maintenance of soil quality and biodiversity through diversified crop and livestock production systems;[56] the use of diverse cultivars; the integration of diverse livestock species, breeds, and age structures within livestock breeds; composting and recycling waste; water conservation practices[57] and alternative energy production;[58] and the use of multiple energy, nutrient and water sources are examples of sustainable agriculture practices that cultivate agroecosystem response diversity.

All of the producers featured in this book manage diversified production systems that exhibit high functional and response diversity to changing ecological and social conditions. Several producers use management strategies that emphasize exceptionally diverse species assemblages, for example, cover crop cocktails (Brendon Rockey, Gail Fuller and Gabe Brown), fruit crop species and cultivars (Steve Ela and Kole Tonnemaker) and crop rotations that feature alternative cultivars or species in different phases of the rotation (Nash Huber and Russ Zenner).

Spatial and temporal diversity support ecological self-regulation, functional diversity and response diversity by providing opportunities for the formation of diverse biological relationships and linkages across space and time. Agroecosystems with high spatial and temporal diversity have a number of distinctly different small-scale ecosystems present in production fields, pastures, edges and natural areas that change over time. This landscape mosaic features patches of both intensively managed and unmanaged land integrated with livestock and wildlife.

All of the producers featured in this book manage production systems that promote spatial and temporal diversity, but several producers emphasized landscape diversity. Paul Muller describes his farm as a diverse mosaic of annual vegetable crops, perennial bushes, tree crops and natural riparian areas. Kole Tonnemaker manages a diverse landscape mosaic of fruit trees, annual crops, hay and grains. Gary Price manages a diverse landscape of improved and native perennial grasslands interspersed with natural lakes and small woodlands. All the featured producers recognize the contribution diversity makes to general resilience, and many view their management of temporal and/or spatial crop and livestock diversity as the best production insurance available to them.

Appropriate connectedness invites a focus on the number and quality of ecological and social relationships within the agroecosystem and between the agroecosystem and the surrounding community. Appropriately connected agroecosystems will build relationships that enhance functional and response diversity. Many weak (i.e., not critical to function) connections are favored over a few strong (i.e., critical to function) connections. Agroecosystems that rely on a few strong connections for critical resources reduce their resilience to events that disrupt those connections; in contrast, many weak connections enhance response capacity.

Agroecosystems managed for modularity and response diversity enhance appropriate connectedness. For example, the modular design of Tom Trantham's pasture-based dairy system reduces the potential for total crop failure because he relies on a diversity of short-season annual forages that are largely independent of each other: the failure of one crop does not influence the success of the rest. In contrast, corn production in an industrial dairy system depends on the availability of fertilizers and pesticides, creating a strong dependence on the agricultural input sector, and total crop failure in this system would require the purchase of replacement feed resulting in a large increase in production costs.

Carefully exposed to disturbance encourages the manager to remember that agroecosystems require disturbance to cultivate response capacity. Frequent, small-scale disturbances that do not exceed the coping capacity of the agroecosystem enhance response and recovery capacity through the release of resources and enables the selection of robust plants, animals, production practices and system configurations and the emergence of novel relationships. Sustainable practices like crop rotation, intensive grazing and managing community dynamics for pest suppression all provide opportunities for frequent, small-scale disturbance. Adaptive management offers a useful framework for learning from the observation of agroecosystem response to disturbance.

All the featured producers use sustainable practices that create opportunities for regular disturbance of the agroecosystem; however, those using holistic or planned grazing strategies (e.g., Julia Stafford, Mark Frasier and Gary Price), integrated pest management strategies (Steve Ela, Kole Tonnemaker, Jim Koan and Jonathan Bishop), plant selection (Nash Huber and Bob Quinn) or careful observation of the growth and development of specific species (Gabe Brown, Brendon Rockey and Gail Fuller) offer the best examples of the use of disturbance to cultivate response capacity.

Coupled with local natural capital describes the capacity of the agroecosystem to function largely within the local carrying capacity. The ability to maintain productivity with local resources increases the modularity of the agroecosystem and reduces the degradation of other regions through the import of natural resources or export of wastes.

Agroecosystems that depend on local natural capital must emphasize the sustainable use and regeneration of local natural resources to remain productive and profitable.

Most of the producers featured in this book offer examples of agroecosystems coupled with local natural capital. Dryland farmers Gail Fuller, Gabe Brown and Russ Zenner and livestock producers Mark Fraser, Gary Price, Will Harris, Ron Rosman and Greg Gunthorp all produce grain and livestock using primarily the local natural capital generated on their farms and ranches, plus natural precipitation. Even the producers of vegetables and fruits—crops that often require intensive fertility, pest management and supplemental irrigation to produce quality crops—manage soil quality, crop diversity and water resources to reduce the need for inputs of fertilizer, pesticides and irrigation.

Social Response Capacity

Social criteria recognize the distinct role humans play in designing and managing agricultural ecosystems and invite a focus on the potential benefits of managing social assets to enhance diversity, modularity and balancing feedbacks in the agroecosystem. The seven proposed social criteria are well aligned with the principles of social and economic sustainability: self-organization, human capital, reflective and shared learning, honors legacy, globally autonomous and locally interdependent, and reasonably profitable.

In *self-organizing social systems*, the people in the agroecosystem are free to organize and engage in productive relationships and site-specific innovation both within and outside the system. Self-organizing agroecosystems exhibit social self-regulation supported by designs that have the capacity to cultivate diversity in and accumulate the human, social, physical and financial capital crucial to their function. Sustainable producers commonly self-organize into grassroots networks, formal and informal institutions such as cooperatives, farmers markets, community supported agriculture, community processing facilities and research and advisory networks.

The cultivation of *human capital* emphasizes human resource development to enhance the resilience of the agroecosystem. This capital can include constructed (economic activity, technology, infrastructure), cultural (individual skills and abilities) and social (social organizations, norms, formal and informal networks) elements. Multiple benefits to farm and community are created by high levels of human capital created through positive educational, economic and social relationships. Investment in infrastructure and institutions to support community-based education, research and development, locally-owned businesses and support for social events in farming communities all serve to enhance human capital assets.

All of the producers featured in this book offer examples of practices that cultivate human capital. Many are active in community-based educational programs. Featured farmers have founded marketing cooperatives (Jim Crawford/Tuscarora Organic, Russ Zenner/Shepard's Grain, Bob Quinn/Montana Flour and Grains) and research cooperatives (Ron Rosmann/Practical Farmers of Iowa). Many have invested in local processing on their farm or in their communities (Bob Quinn, Greg Gunthorp, Will Harris, Tom Trantham, Jim Bishop, Jim Koan), while others have invested in retail stores on their farms (Tom Trantham, Jim Bishop, Ron Rosmann, Nash Huber). All of these activities enhance the self-organizing capacity of the food systems within which the featured agroecosystems reside and so serve to regenerate and accumulate a diversity of human, social, financial and physical capital.

Reflective and shared learning recognizes the importance of blending local knowledge with scientific knowledge to develop place-based innovations that enhance agroecosystem resilience. Practices that support reflective and shared learning in agroecosystems include adaptive management strategies, preservation of historical records and participation in producer-based research and education networks. All of the producers featured in this book regularly engage in community-based activities that support reflective and shared learning. Most of the livestock producers use intensive grazing practices, and many practice holistic management. Featured producers have founded research cooperatives (Ron Rosmann/Practical Farmers of Iowa), participated in collaborative research on their farms (Russ Zenner, Ron Rosmann, Gabe Brown, Greg Gunthorp, Tom Trantham, Gary Price) or conducted formal research on their own (Nash Huber, Bob Quinn).

Honors legacy recognizes the powerful influence that past conditions and experiences have on present identity and future evolution of the agroecosystem. The biological and cultural memory embodied in an agroecosystem is a rich source of place-based knowledge and experience that can be drawn on for inspiration and innovation. The maintenance of heirloom seeds, engagement with elders and the exploration of traditional cultivation and food-processing techniques are examples of activities that preserve, honor and celebrate the cultural and ecological heritage of place. For example, Jim Koan's Original Hard Cider has been made on his farm for well over a hundred years. Jim uses the same apple varieties and the same recipe as his great-great grandfather did back in the 1850s.

Global autonomy and local interdependence invites attention to the value of local relationships to buffer the agroecosystem from global conditions. A locally interdependent agroecosystem is less vulnerable to disturbances caused by social conditions that are outside the influence or control of the producer. Sustainable agriculture practices that support global autonomy and local interdependence include direct marketing to local and regional markets, reduced reliance on external inputs and participation in local purchasing and marketing cooperatives. All of the featured producers offer models of globally autonomous and locally interdependent agroecosystems; however, a few sell into commodity markets (Julia Stafford, Mark Fraser, Gabe Brown and Gail Fuller) and several others sell value-added commodities into regional (Russ Zenner, Ron Rosmann and Will Harris), national (Russ Zenner and Bob Quinn) or global markets (Bob Quinn). All of the featured agroecosystems are dependent on energy inputs—electricity, gasoline and diesel fuel—that are largely outside the influence and control of the producer.

Reasonably profitable focuses attention on the financial assets produced by the agroecosystem. The production of profit supports the well-being of the people working in the agroecosystem and builds the wealth required to support response, recovery and transformation capacity. Reasonably profitable agroecosystems support the well-being of the people that work in them by providing a living wage to labor, a reasonable return on invested capital and the financial resources needed for health care, education, family activities, retirement and other desired activities and services.

All of the featured producers manage resilient and sustainable agroecosystems that have remained remarkably productive and profitable, largely without the benefit of the government support enjoyed by industrial agriculture, over many decades of dynamic change in American agriculture. All of the featured agroecosystems fully employ a number of business partners and/or family members and many provide well-paying full-time, part-time and seasonal employment to permanent residents in their communities.

Considered together, these thirteen proposed resilience criteria fully address the ecological, economic and social principles of sustainable agriculture in the requirements for site-specific, ecosystem-based design; the emphasis on managing ecosystem processes through planned biodiversity; the focus on the economic well-being of producers, agricultural workers and society as a whole; and the recognition of linkages between the agroecosystem and community. The criteria also extend sustainability principles by inviting a focus on redundancy in critical system functions and the identification and development of robust system components through intentional disturbance.

Although there is still much to learn about the principles and practices of a resilient agriculture, this set of resilience criteria builds upon the well-established principles and practices of sustainable agriculture to offer a useful new framework for the design and assessment of agroecosystems; however, resilience concepts have also been applied to the development of effective strategies for agricultural adaptation to climate change.

Adapting Agriculture to Climate Change: Resistance, Resilience and Transformation

Agricultural adaptation strategies can be classified as falling along a resistance, resilience and transformation continuum that describes different management objectives.[59] *Resistance strategies* are actions that protect the existing agroecosystem from climate effects, while *resilience strategies* improve the capacity of the existing agroecosystem to cope with and recover from climate-related stress, disturbances and shocks. *Transformation strategies* are actions that facilitate the transition of the existing agroecosystem to a new identity that is more resilient to current or projected climate conditions. Resilient systems typically exhibit some characteristics of all three kinds of management strategies.[60] Taken as a whole, this range of management approaches represents the beginnings of a climate resilience toolkit for US farmers and ranchers.

Enhancing Adaptive Capacity: Resistance Strategies

Resistance strategies are actions taken to protect the existing agroecosystem from damaging climate change effects without altering its identity. They emphasize the use of physical and financial assets rather than natural, human and social assets. Investments in physical assets protect specific components of the system from specific climate effects, while financial assets aid recovery when the physical protection fails. Resistance strategies can be costly, will likely increase in cost and decrease in effectiveness over time and may ultimately fail if the pace and intensity of climate change increase as projected.[61]

Some key resistance strategies in use by US producers involve changes in equipment and infrastructure. For example, corn and soybean producers in the Midwest are increasingly challenged by more variable and shorter windows in which to complete the seasonal fieldwork required for crop production. These farmers have responded to this increased uncertainty by purchasing larger or additional equipment (tractors and associated implements) to reduce the time needed to complete fieldwork and increase the management flexibility of their farm systems.[62] Many Midwest farmers are also adding or upgrading infrastructure such as irrigation, drainage and levee systems to improve defenses against drought and more frequent heavy rainfall.

Federally funded farm support programs encourage the use of resistance strategies by producers growing eligible crops, primarily commodity crops like corn, wheat, soybeans and rice. These programs offer subsidized insurance against production risks, cost-share programs for infrastructure improvements and disaster relief. Other practices that can aid resistance strategies include planning for likely weather-related hazards through the use of weekly and seasonal weather forecasts, the use of pest and disease forecasts, and increasing on-farm reserves of critical materials (e.g., water, fuel, feed) that would be difficult to obtain during a catastrophic weather-related disturbance. Recent regional climate history and regional climate projections may be generally useful in longer-term planning, for example, when investing in infrastructure with a life expectancy of fifteen to twenty-five years. In the last five years, federal and state government efforts to support effective climate change adaptation in agriculture and other sectors of the economy have accelerated.[63]

Like industrial producers throughout the United States, the sustainable farmers and ranchers featured in this book employ a number of resistance practices. Nash Huber purchased additional equipment to speed the completion of his fieldwork. Paul Muller is considering upgrading to a more efficient irrigation system. Jacquie Monroe has expanded irrigated production area and upgraded to more efficient irrigation systems. Jim Crawford physically protects his fields from excessive spring rainfall with plastic mulch and has increased pesticide use. Steve Ela purchased additional wind machines to enhance frost protection in his orchard. Jim Koan increased the capacity of recently installed field drainage systems. Jim Hayes installed new drainage and built new sheltered space for livestock.

Resistance strategies have an important role to play in efforts to enhance agroecosystem resilience, particularly in defending high-value crops near the end of long stages of development or in short stages of high productivity (e.g., tree crops) and/or crops that can only be grown in limited areas, such as those that require a Mediterranean environment (e.g., almonds). Resistance strategies can also play an important short-term role during the transition to more resilient production systems; however, a focus on resistance as a climate risk management strategy can increase the risk of maladaptation. Maladaptation is the investment in short-term resistance strategies at the expense of investments in resilience and transformation strategies that, over the long term, may be less costly and have more potential to sustain productivity in a changing climate.[64]

9.1. Federal Subsidies Promote Maladaptation

For more than seventy years, federal farm subsidies have encouraged the industrialization of the US food system through programs that favor the industrial production of specific crops and livestock. Farm subsidy programs were designed to stabilize agricultural production and prices through the use of direct payments and subsidized insurance. Federally subsidized insurance programs have increased in importance relative to direct payment programs since the mid-1980s. The Agricultural Act of 2014, more commonly referred to as the 2014 Farm Bill, continues this trend with a major shift in federal support from direct payment programs to insurance programs as the primary safety net for US farmers.

Weather has always been an important factor in agricultural production, but climate risk—the damages caused by the increased weather variability and extreme weather events associated with climate change—has emerged as a novel hazard that is projected to increase in intensity through this century. Publicly subsidized programs that insure against climate risk without requiring investments to enhance adaptive capacity can act as barriers to effective adaptation in agriculture.[65]

The 2014 Farm Bill continues a primary focus on the support of industrial commodity production and disaster relief programs.[66] The practice of offering unlimited subsidies to producers, landowners and farm investors regardless of their income or ability to pay will also continue. This lack of means testing encourages the expansion of large, investor-owned agricultural operations, drives up land prices and makes it increasingly difficult for young and beginning farmers to enter the business. Annual insurance program costs are projected to average about nine billion dollars per year over the next decade.

This maladaptive path squanders financial resources that could be used to cultivate a sustainable and resilient US agriculture in an effort to protect the existing industrial system; however, there is an alternative. Federal farm subsidies could be redirected away from industrial agriculture and toward a more sustainable and resilient agriculture with the capacity to produce multiple benefits to society while sustaining the American food supply in changing climate conditions.

Fortunately, the 2014 Farm Bill does include some changes that put US agriculture on a more resilient path.[67] New and redesigned insurance programs increase crop insurance options for pasture-based livestock producers, more diversified production systems and organic producers; offer increased subsidy to new and beginning farmers; and link eligibility for federal subsidies to improved soil management for some producers. These kinds of changes to federally subsidized insurance programs are important first steps toward the major changes in farm support programs needed to support farmers and ranchers who produce food in sustainable and resilient produc-

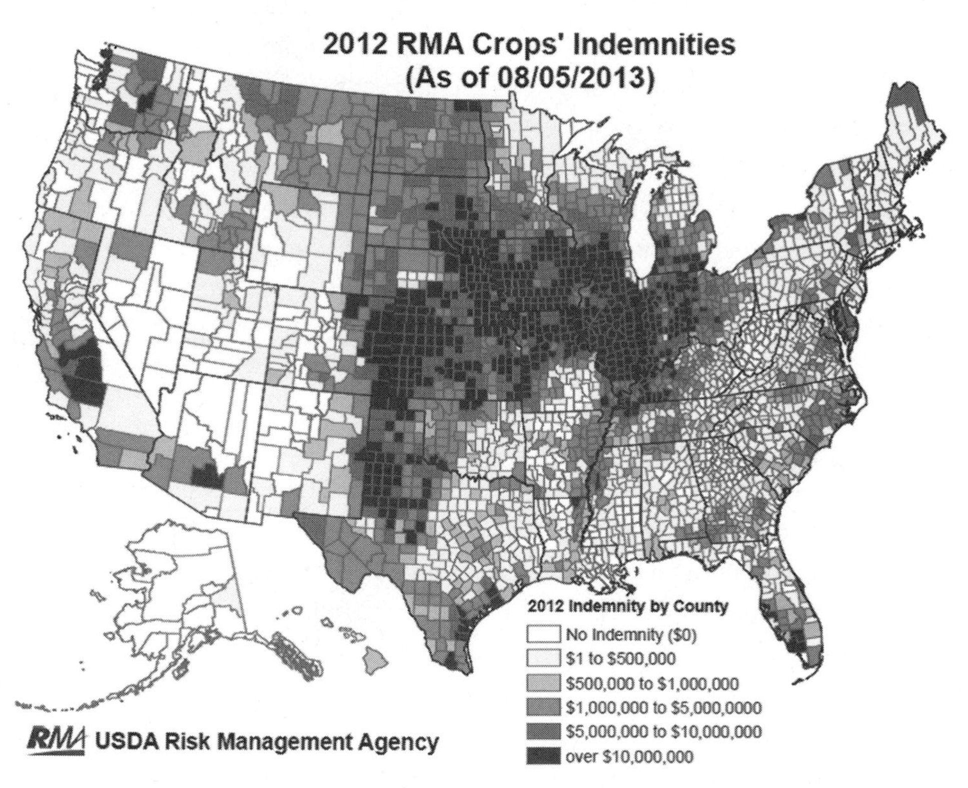

FIGURE 9.2. RMA Indemnity Payments to Crop Producers Tops 14 Billion in 2012. The 2012 drought was more extensive than any since the 1950s, with about 80 percent of agricultural land in the United States in drought conditions. Indemnity payments in 2012 totaled 14 billion dollars, with 90 percent going to just four crops: corn, soybeans, cotton and wheat. Projected increases in the frequency and intensity of extreme weather will increase the costs of federally-subsidized insurance programs without a redirection of subsidies to promote more sustainable and resilient agricultural production systems. Credit: USDA

tion systems, but much more could be done. For example, no new legislation would be needed to offer reduced insurance premiums to farmers who use proven techniques such as cover crops, conservation tillage and other soil health-building management practices to improve farm resilience to climate change.[68]

Enhancing Adaptive Capacity: Resilience Strategies

Resilience strategies are actions taken to enhance the response and recovery capacity of the agroecosystem.[69] These actions may involve some changes to the structure and function of the system, but not so much as to change the essential identity of the system. The producer designs for and manages the response capacity of the agroecosystem and may intervene in some cases to prepare for or respond to a climate disturbance; however, minimal management intervention is the desired goal. Resilience strategies cultivate the full range of capital assets—human, natural, and social, financial and physical—to enhance the adaptive capacity of the agroecosystem. Like resistance strategies, they may increase in cost and decrease in effectiveness as climate change proceeds and will fail if the response capacity of the agroecosystem cannot keep pace with changing climate conditions.

Resilience strategies employed by US industrial producers include the use of drought-resistant species and crop varieties and changes in management to promote soil quality, such as conservation tillage, the use of cover crops and other practices that increase the quantity and quality of soil organic matter.[70] Industrial corn and soybean producers also recognize the climate risk benefits of more diversified cropping systems and consider diversification to be a key adaptation strategy if climate change effects intensify, along with purchasing more crop insurance (a resistance strategy) and going out of business (a transformation strategy).[71] Recommended practices for livestock producers managing for resilience to climate change include obtaining the education and training required to understand and manage animal needs, potential stress levels and options for reducing stress, and selecting animals and management strategies compatible with local production conditions.[72]

All of the sustainable producers featured in this book appreciate the resilience benefits offered by high-quality soils, diversified production systems, value-added processing and direct marketing. Other resilient adaptations include: reorienting production windows to avoid extreme weather conditions; increasing crop diversity and flexibility of diversified crop rotations; increasing investment in recovery reserves to buffer more variable weather conditions; and capturing new opportunities as average temperatures increase and the growing season lengthens.

Nash Huber and Russ Zenner have increased crop production flexibility by identifying a range of crop types adapted to different weather conditions for each phase of their crop rotations. Gabe Brown and Gail Fuller have developed incredibly diverse cover crop cocktails to assure good cover crop production across variable soil, microclimate and weather conditions. Tom Trantham and Gail Fuller manage an annual/perennial intercrop with a "pasture cropping" system that features the no-till planting of forages into permanent pasture. Nash Huber and Bob Quinn are screening plant species and adapting crop cultivars robust to local conditions and more variable weather and extremes. Steve Ela has matched crop species and cultivars to microclimate variations across his orchard landscape. Gabe Brown, Will Harris and Greg Gunthorp are actively selecting for livestock that perform well in their agroecosystems under more variable conditions.

Jim Koan has integrated livestock into his orchard systems, and Jacquie Monroe has integrated livestock, feed and forage production into her vegetable production system. Kole Tonnemaker and Steve Ela increased product diversity (adding annuals to perennial cropping systems) to spread production risks in more variable weather, to take advantage of longer growing seasons and to respond to consumer demands in direct markets. Alex Hitt is shifting his crop production away from the increasingly intense heat and drought of mid-summer. Both Ken Dawson and Alex Hitt have made changes in planting schedules to avoid pest damages and interference with pollination. Elizabeth Henderson installed wells and irrigation in response to increasingly frequent and intense summer droughts. Jim Hayes installed solar panels to provide backup power to freezers in the event of a loss of power.

Livestock producers Julia Stafford, Mark Frasier, Gary Price, Gabe Brown, Gail Fuller, Dick Cates, Ron Rosmann, Jim Hayes, Tom Trantham and Will Harris enhance ecological response capacity by managing herd size and movement through improved pastures, annual forage crops and/or native grasslands using intensive grazing and holistic management strategies. Many have reduced herd size, leased additional grazing acreage or done both in response to more variable weather and extremes. Greg Gunthorp uses standing corn as a sheltered area for pigs and encourages mulberry trees in wooded field edges for shelter and supplemental feed.

These resilient adaptive actions are examples of the many opportunities available in sustainable agroecosystems to cultivate resilience to more variable weather and extremes through enhanced ecological and social response capacity.

Enhancing Adaptive Capacity: Transformation Strategies

Transformation strategies are actions that facilitate the transition of the existing agroecosystem to a new identity that is more resilient to current or future climate conditions. These strategies enhance adaptive capacity by changing the fundamental structure, function and purpose—the identity—of the existing agroecosystem. These kinds of changes could involve increasing the diversity of crops grown on the farm or shifting from housed to pasture-based livestock management. They are design and management changes that alter the relationships among the crops, livestock and land so that ecosystem functions such as energy flow, water and nutrient cycles and community dynamics are better adapted to local ecosystems under changing climate conditions.

The time to consider the transformation of an agroecosystem is when the system, in its current configuration, fails to achieve its purpose; when the costs (ecological, social, economic) of maintaining it begin to increase or become unacceptable; or when it becomes clear that a transition is inevitable because of changing capital assets. The capacity of the system to transition to a new identity is its transformative capacity. Transformative capacity, like response and recovery capacity, is promoted by high levels of all five capital assets; however, it also requires two other key qualities: the recognition of the need for transformative change and the ability to envision and access practical and credible options for change. Recall that adaptive capacity is a function of the operating context, the individual capability to act and existing knowledge and options. High adaptive capacity is associated with high transformative capacity.

Sustainable agricultural practices offer a wealth of practical and credible options for the transformation of agroecosystems. Depending on the configuration of the existing system, sustainable agricultural principles and practices offer a diversity of options. The transition to diversified annual crop rotations; the intercropping of annual and perennial crops; transitioning from annual to perennial crops; integrating of livestock into crop production systems; intensive grazing management (also called planned grazing, mob grazing and mob stocking); converting cropland to pasture, agroforestry and renaturalized areas; and native grasslands or wetlands restoration are examples of practices that change the identity of the agroecosystem to enhance resilience to climate change.[73]

Sustainable agriculture practices that offer transformative options for livestock production systems may include a transition to livestock species or breeds that have greater tolerance of changing climate conditions, forages and grassland quality and pests and diseases; and the use of multispecies production systems (for example, following cattle with poultry). The conversion from industrial livestock production in confined animal feeding operations to integrated crop/livestock production systems is a transformative strategy that has been recommended to reduce detrimental environmental impacts, improve the profitability and sustainability, and enhance resilience to climate change in US livestock production systems.[74]

Some of the producers featured in this book offer examples of the power of transformative change. *Brendon Rockey* transformed an industrial potato production system by restructuring crop diversity to enhance soil quality and water conservation. *Gabe Brown* used holistic management, an adaptive management strategy, to guide the transformation of an industrial grain production system and degraded native prairie into a sustainable and resilient diversified crop and livestock production system. With a focus on building soil quality and biodiversity, Gabe diversified crop rotations with cash crops and cover crop cocktails, integrated livestock into the crop rotation and restored the native prairie through management-intensive grazing practices. *Gail Fuller* used holistic management to transition his industrial grain farm into a diversified integrated crop and livestock production system with a focus on building soil quality with cover crop cocktails and management-intensive grazing. *Greg Gunthorp* transformed a pastured hog production system into a diversified livestock production system with an on-farm processing plant and retail store. *Will Harris* transformed an industrial cattle feeding system into a certified organic, diversified, pasture-based livestock production system with an on-farm processing plant and retail store.

From Farm Gate to Plate: Barriers to the Cultivation of a Resilient Agriculture

Both theory and experience suggest that managing for resilience at one scale often requires changes at scales both below and above the focal scale. In other words, the operating context of a farm or ranch offers both opportunities and barriers to cultivating resilience. Sometimes the challenges are most apparent in components that are under the producer's control, but most often challenges to cultivating adaptive capacity emerge at the scale just above farm or ranch scale.[75] As an example, consider this comparison of the challenges encountered during the transition to sustainable and resilient agroecosystems at Rockey Farms and Fuller Farms.

Brendon Rockey's transition from an industrial model (emphasizing financial and physical assets) to a more sustainable and resilient model (emphasizing natural, social and human assets) required a transformation of soil quality, crop diversity and management perspective—all components of the agroecosystem under Brendon's management. He recognized that the agroecosystem had to change (water supply was dwindling), was aware of practical and credible alternatives (soil building cover crops) and had access to all the assets he needed to successfully guide the transition. The barriers presented by the operating context were primarily within the agroecosystem, although Brendon says he did have to put up with a lot of ridicule from his neighbors and friends during the early stages of the transition.

Gail Fuller's transition to a sustainable and resilient production system has been hampered by a lack of information about crop options and conflicting and unreasonably restrictive federal crop insurance program regulations. Gail recognized the need for change (soil quality degradation) and was aware of practical and credible alternatives developed in other regions (cover crop cocktails and increased diversity of cash crops), but encountered barriers at the scale above the farm level: agricultural advisory services could not provide guidance, and conflicting federal program guidance restricted Gail's ability to innovate and maintain the recovery capacity provided by subsidized crop insurance. These barriers created significant challenges, and both continue to challenge Gail's efforts to develop a more sustainable and resilient dryland agroecosystem at Fuller Farms.

An essential component of Jim Koan's sustainable and resilient fruit production system (pigs in the orchard) was threatened by proposed federal crop handling and food safety regulations created to address the food safety risks inherent in industrial production systems. These threats emerged in the operating context of Almar Orchards at the scale above the farm scale. Jim recognized the need for change (the current farm system did not meet proposed regulations), discovered a credible and practical alternative (hard cider) and had access to all the assets he needed to achieve the transition. In this case, the agroecosystem had sufficient adaptive capacity to innovate a solution in response to threats that were not under the control of the producer.

This chapter and Chapter 4 present the kinds of changes in agroecosystem components that enhance the sustainability and resilience of American farms and ranches. These changes are largely under the control of the producer, subject to the barriers and opportunities presented by the operating context. Although many US producers perceive a need for a change (*recognize need*) and this book, and many others, offers a variety of "proof of concept" models that are well adapted to the diversity of American landscapes (*awareness of credible and practical alternatives*), the operating context of American agriculture creates barriers to the capacity of farmers and ranchers to self-organize, learn, innovate, adapt and transform their agroecosystems in response to changing climate conditions.

Managing for sustainability and resilience objectives at the individual farm and ranch scale has been constrained by government and corporate actions that have driven the industrialization and globalization of the US food system since the 1850s. These forces strive to maintain the existing system despite the fact that it has not fulfilled its purpose (save the family farm, feed the world and provide Americans with a safe, abundant and inexpensive food supply), and there is overwhelming evidence that it is not sustainable or resilient to twenty-first-century challenges. Cultivating a resilient agriculture cannot be accomplished through actions taken only at the individual farm or ranch scale. A resilient agriculture will require a transformation of the operating context of American agriculture—the US industrial food system.

Endnotes

1. National Research Council. A Pivotal Time in Agriculture. Ch. 1 in *Toward Sustainable Agricultural Systems in the 21st Century*. 2010. National Academies Press, Washington, DC.
2. National Research Council. Understanding Agricultural Sustainability. Ch. 1 in *Toward Sustainable Agricultural Systems in the 21st Century*. 2010. National Academies Press, Washington, DC.
3. R. T. Munang, I. Thiaw, and M. Rivington. 2011. Ecosystem management: Tomorrow's approach to enhancing food security under a changing climate. *Sustainability*, 3(7): 937–954.
4. World Bank. 2009. Convenient solutions to an inconvenient truth: Ecosystem-based approaches to climate change. climate-l.org/2009/07/06/world-bank-publishes-report-on-ecosystem-based-approaches-to-climate-change/
5. Ecosystem-based Approaches to Climate Change Adaptation and Mitigation in Germany, Austria and Switzerland. ecologic.eu/7544
6. R. Munang et al. 2013. Climate Change and Ecosystem-based Adaptation: A New Pragmatic Approach to Buffering Climate Change Impacts. *Current Opinion in Environmental Sustainability*, Vol. 5:1–5.
7. J. Liu et al. 2011. *Restoring the natural foundation to sustain a green economy: A century-long journey for ecosystem management.* International Ecosystem

Management Partnership, United Nations Environmental Program Policy Brief 6. unep.org/ecosystemmanagement/Portals/7/Documents/policy%20 series%206.pdf.
8. S. Polefka. 2014. Coastal Ecosystem Protection: A Strategic Opportunity for the United States and India.
9. G. Poppy et al. 2014. Achieving food and environmental security: New approaches to close the gap. Phil. Trans. R. Soc. B 369:20120272. dx.doi.org /10.1098/rstb.2012.0272.
10. International Union for the Conservation of Nature. nd. Protected Areas Helping People Cope with Climate Change. World Commission on Protected Areas. iucn.org/wcpa
11. S. Scherr and J. Neely. 2008. Biodiversity Conservation and Agricultural Sustainability: Towards a New Paradigm of Ecoagricultural Landscapes. Phil. Trans Royal Society, Vol. 363: 477–494.
12. C. Walthall et al. Adapting to Climate Change. Ch. 6 in *Climate Change and Agriculture in the United States: Effects and Adaptation*. 2012. USDA Technical Bulletin 1935.
13. C. Walthall et al. Managing Climate Risk: New Strategies for Novel Uncertainty, Adapting to Climate Change. Ch. 7 in *Climate Change and Agriculture in the United States: Effects and Adaptation*. 2012. USDA Technical Bulletin 1935.
14. K. Moore. 2009. Landscape Systems Framework for Adaptive Management. Ch. 1 in *The Sciences and Art of Adaptive Management: Innovating for Sustainable Agriculture and Natural Resource Management*. Ankeny, IA: Soil and Water Conservation Society.
15. C. Walthall et al. Adapting to Climate Change. Ch. 6 in *Climate Change and Agriculture in the United States: Effects and Adaptation*. 2012. USDA Technical Bulletin 1935.
16. R. Janke. 2000. Whole Farm Planning for Economic and Environmental Sustainability. Kansas State University Agricultural Experiment Station and Cooperative Extension Service. Publication MF-2403.
17. For example, see L. Lengnick and S. Kask. 2009. Helping Farmers Make Complex Choices: A Sustainable Decision Tool for Farmers. researchgate. net; and E. Henderson and K. North. 2011. Whole Farm Planning: Ecological Imperatives, Personal Values and Economics, Second Edition. Northeast Organic Farming Association (NOFA) Interstate Council, Organic Principles and Practices Handbook Series.
18. K. Mackenzie and L. Kemp. 1999. Whole Farm Planning at Work: Success Stories of 10 Farms. The Minnesota Project. Minnesota Institute for Sustainable Agriculture.

19. R. Janke. 2000. Whole Farm Planning for Economic and Environmental Sustainability. Kansas State University Agricultural Experiment Station and Cooperative Extension Service. Publication MF-2403.
20. For example, see Organic Whole Farm Planning, Ohio Ecological Food and Farming Association; Whole Farm Planning for Beginners, Virginia Cooperative Extension; Plan and Manage the Whole Farm, North Carolina Cooperative Extension; Farm Beginnings Program, Nebraska Cooperative Extension; Holistic Management: A Whole Farm Decision-Making Framework, ATTRA.
21. National Research Council. Understanding Agricultural Sustainability. Ch. 1 in *Toward Sustainable Agricultural Systems in the 21st Century*. 2010. National Academies Press, Washington, DC.
22. For example: State, Local, and Tribal Leaders Task Force on Climate Preparedness and Resilience. 2013. whitehouse.gov/administration/eop/ceq/initiatives/resilience/taskforce; Resilient Communities for America. 2014. icleiusa.org/library/documents/rc4a-federal-policy-initiative-report-5.12.14; Weathering the Storm: Building Business Resilience to Climate Change. 2013. Center for Climate and Energy Solutions; Resilient Cities Series. 2014. Annual Global Forum on Resilience and Adaptation. ICLEI-Local Governments for Sustainability.
23. B. Walker and D. Salt. 2012. Practicing Resilience in Different Ways. Ch. 5 in *Resilience Practice: Building Capacity to Absorb Disturbance and Maintain Function*. Island Press: Washington, DC.
24. A person, a farm, a community and an ecosystem are all examples of complex adaptive systems. Complex adaptive systems are defined by three requirements: 1. they are made up of independent and interacting components—for example, the species found on a farm; 2. the components are subject to forces that serve to promote or resist change in the system—for example, the population increase of a pest species and the predation of that pest by another species; 3. the system produces constant variation and novelty through changes in existing components or the entry of new components to the system—for example, the changes that accompany plant growth and development or the addition of new species in crop rotation.
25. National Research Council. Understanding Agricultural Sustainability. Ch. 1 in *Toward Sustainable Agricultural Systems in the 21st Century*. 2010. National Academies Press, Washington, DC.
26. D. Nelson et al. Adaptation to Environmental Change: Contributions of a Resilience Framework. 2007. *Annu. Rev. Environ. Resour*, 32:395–419.
27. Climate Change 2007: Impacts, Adaptation and Vulnerability. p. 880. Contribution of Working Group II to the Fourth Assessment Report of

the Intergovernmental Panel on Climate Change, M. L. Parry, O. F. Canziani, J. P. Palutikof, P. J. van der Linden and C. E. Hanson, eds., Cambridge University Press, Cambridge, UK.

28. See chapter 4.
29. This discussion of the qualities of resilient systems is adapted largely from B. Walker and D. Salt. 2012. The Essence of Resilience Thinking. Ch. 1 in *Resilience Practice: Building Capacity to Absorb Disturbance and Maintain Function*. Island Press: Washington, DC.
30. B. Walsh. 2013. 10 Years After the Great Blackout, the Grid Is Stronger—but Vulnerable to Extreme Weather. *Time*, Aug. 13. science.time.com/2013/08/13/ten-years-after-the-great-blackout-the-grid-is-stronger-but-vulnerable-to-extreme-weather/
31. C. Minoiu and S. Sharma. 2014. Financial Networks Key to Understanding Systemic Risk. *IMF Survey Magazine*. imf.org/external/pubs/ft/survey/so/2014/RES052314A.htm
32. W. Fry et al. 2009. The 2009 Late Blight Pandemic in Eastern USA. The American Phytopathological Society. apsnet.org/publications/apsnet features/Pages/2009LateBlight.aspx
33. S. J. Goerner, B. Lietaer, and R. E. Ulanowicz. 2009. Quantifying economic sustainability: Implications for free-enterprise theory, policy and practice. *Ecological Economics*, 69:76–81.
34. D. Meadows. 2008. Leverage Points: Places to Intervene in a System. In *Thinking in Systems: A Primer*. Earthscan: New York.
35. Ibid.
36. Ibid.
37. F. Magdoff and H. van Es. 2009. The soil degradation spiral, Fig. 1.1. p. 5, in Building Soils for Better Crops. Sustainable Agriculture Research and Education Program: Washington, DC.
38. D. Meadows. 2008. Leverage Points: Places to Intervene in a System. In *Thinking in Systems: A Primer*. Earthscan: New York.
39. Ibid.
40. Ibid.
41. S.J Goerner, B. Lietaer, and R. E. Ulanowicz. 2009. Quantifying economic sustainability: Implications for free-enterprise theory, policy and practice. *Ecological Economics*, 69:76–81.
42. See chapter 4 discussion of capital assets and adaptive capacity.
43. S. J. Goerner, B. Lietaer, and R. E. Ulanowicz. 2009. Quantifying economic sustainability: Implications for free-enterprise theory, policy and practice. *Ecological Economics*, 69:76–81.

44. A full discussion of these considerations is beyond the scope of this book. An accessible and more detailed discussion of resilience principles and practices in landscape scale management can be found in B. Walker and D. Salt. 2012. *Resilience Practice: Building Capacity to Absorb Disturbance and Maintain Function*. Island Press: Washington, DC.
45. For example, see The EcoTipping Points Project: Models of Success in a Time of Crisis. ecotippingpoints.org/index.html
46. D. Meadows. 2008. Leverage Points: Places to Intervene in a System. In *Thinking in Systems: A Primer*. Earthscan.
47. B. Walker and D. Salt. 2012. Assessing Resilience. Ch. 3 in *Resilience Practice: Building Capacity to Absorb Disturbance and Maintain Function*. Island Press: Washington, DC.
48. See chapter 4 discussion of adaptive capacity as a component of vulnerability.
49. B. Walker and D. Salt. 2012. Assessing Resilience. Ch. 3 in *Resilience Practice: Building Capacity to Absorb Disturbance and Maintain Function*. Island Press: Washington, DC.
50. G. Berardi, R. Green, and B. Hammon. 2011. Stability, sustainability, and catastrophe: Applying resilience thinking to U. S. agriculture. *Human Ecology Review*, Vol. 18 (2): 115–125.
51. The natural, human, social, physical and financial resources available to the farm manager—see chapter 4.
52. For example, A Guide to the Art & Science of On-Farm Monitoring: The Monitoring Tool Box. Land Stewardship Association: Minneapolis.
53. J. Cabell and M. Oeofse. 2012. An indicator framework for assessing agroecosystem resilience. *Ecology and Society*, 17:18.
54. See chapter 4.
55. See discussion of the four farming system processes in chapter 4.
56. C. Kremen et al. 2012. Diversified Farming Systems: An Agroecological, Systems-based Alternative to Modern Industrial Agriculture. *Ecology and Society*, 17(4): 44.
57. Smart Water Use on Your Farm or Ranch. 2006. Sustainable Agriculture Network. USDA SARE:Washington, DC. sare.org/content/download/29697/413106/Smart_Water_Use_on_Your_Farm_or_Ranch.pdf
58. Clean Energy Farming: Cutting Costs, Improving Efficiencies, Harnessing Renewables. 2012. Sustainable Agriculture Network. USDA SARE:Washington, DC. sare.org/content/download/3842/36429/Clean_Energy_Farming.pdf?inlinedownload=1
59. C. Walthall et al. Adapting Agriculture to Climate Change. Ch. 7 in *Climate*

Change and Agriculture in the United States: Effects and Adaptation. USDA Bulletin 2935.

60. National Research Council. Understanding Agricultural Sustainability. Ch. 1 in *Toward Sustainable Agricultural Systems in the 21st Century*. 2010. National Academies Press, Washington, DC.
61. C. Walthall et al. Adapting Agriculture to Climate Change. Ch. 7 in *Climate Change and Agriculture in the United States: Effects and Adaptation*. USDA Bulletin 2935.
62. Inid.
63. For example, in 2014 the federal government released the U.S. Climate Resilience Toolkit and USDA initiated the Climate Hubs program. The Climate Resilience Toolkit is an interactive website designed to serve interested citizens, communities, businesses, resource managers, planners and policy leaders at all levels of government by providing scientific tools, information and expertise to help people manage their climate-related risks and opportunities, and improve their resilience to extreme events. USDA's Climate Hubs are designed to develop and deliver science-based, region-specific information and technologies, with USDA agencies and partners, to agricultural and natural resource managers to enable climate-informed decision-making and effective adaptation. toolkit.climate.gov. climatehubs.oce.usda.gov/.
64. C. Walthall et al. Adapting Agriculture to Climate Change. Ch. 7 in *Climate Change and Agriculture in the United States: Effects and Adaptation*. USDA Bulletin 2935.
65. R. McLeman and B. Smit. 2006. Vulnerability to Climate Change Hazards and Risks: Crop and Flood Insurance. *The Canadian Geographer*, 50: 217.
66. National Sustainable Agriculture Coalition. 2014. What Is in the Farm Bill for Sustainable Agriculture and Food Systems? sustainableagriculture.net/blog/2014-farm-bill-outcomes/
67. Ibid.
68. C. O'Connor. 2013. Soil Matters: How the Federal Crop Insurance Program should be reformed to encourage low-risk farming methods with high-reward environmental outcomes. Natural Resources Defense Council: Washington, DC.
69. C. Walthall et al. Adapting Agriculture to Climate Change. Ch. 7 in *Climate Change and Agriculture in the United States: Effects and Adaptation*. USDA Bulletin 2935.
70. Ibid.
71. R. Rejesus et al. 2013. U.S. Agricultural Producer Perceptions of Climate Change. *Journal of Agricultural and Applied Economics*, 45:701–18.

72. C. Walthall et al. Adapting Agriculture to Climate Change. Ch. 7 in *Climate Change and Agriculture in the United States: Effects and Adaptation*. USDA Bulletin 2935.
73. Ibid.
74. Ibid.
75. B. Walker and D. Salt. 2012. *Resilience Practice: Building Capacity to Absorb Disturbance and Maintain Function*. Island Press: Washington, DC.

CHAPTER 10

Cultivating Food Systems for a Changing Climate

"Eating is an agricultural act." This simple statement brings to mind the complex network of connections linking food producer to consumer in the global industrial food system.[1] Global demand for agricultural products and resource availability—labor, land and capital—have shaped the operating context of North American agriculture for more than three hundred years.

Colonial plantation agriculture supplied tobacco, rice, indigo and cotton to Europe in the seventeenth, eighteenth and nineteenth centuries. California's Central Valley supplied grain to Europe and Australia in the nineteenth century. In the twentieth century, a highly concentrated and specialized US food system produced foods and feeds for global markets. No matter the time or place, US industrial agriculture has prospered within an operating context that encouraged the exploitation of labor and natural resources to meet market demand with little regard for the costs to community well-being.

The US industrial food system has proven remarkably resilient over the last hundred and fifty years, adapting to a diversity of production conditions across North America to supply commodities to national and international markets. This resilience has been largely achieved through continuous financial subsidy, public support of research and development to benefit the agricultural industrial complex, and natural and social resource subsidies produced through the degradation of soil, water, air quality, biodiversity, community health and well-being.[2] Paradoxically, public support for this system has at the same time eroded its resilience, because this support shelters the system from stress, disturbances and shocks.[3]

National conditions of climate change exposure, sensitivity and adaptive capacity vary across the United States, so climate vulnerability will likely emerge in regional patterns.[4] Exposures vary widely across the country, as does sensitivity, which is determined by the responses of crop and livestock production systems and associated manufacturing and distribution centers to climate exposures.[5] Adaptive capacity is also place-based and varies with natural resource quality, infrastructure, governance, education, wealth and other social measures of well-being. Because the US food system is tightly connected to global markets, its climate change vulnerability is also determined, in part, by both national and global conditions. Climate change is already disrupting the global food system, and these disturbances are projected to increase in number and intensity in coming years.[6] Global influences such as international agreements, policy and politics, social unrest and environmental changes are both affected by and influence US food system behavior.

Production and consumption vulnerabilities to climate change have received the most attention by agricultural and food security researchers to date, but climate change effects are already being felt throughout the world's food systems by businesses involved in the agricultural supply, processing, distribution and retailing sectors.[7] Global food manufacturers and retailers report a variety of climate-related risks currently under active management, including variability in raw material supplies, lower quality and higher prices, increasing water scarcity, more disruptions and failures of distribution networks, increased incidence of workforce disruption and changing consumer demand.[8] Global changes in ecosystem services, climate disruptions and food security in other countries will influence the price, quality and availability of food to US consumers.[9]

Research suggests that food systems are more vulnerable to global environmental changes when they have the following characteristics: heavy reliance on external or distant resources; low diversity; inequitable access to resources; inflexible governance; highly specialized production, supply and marketing chains; and subsidies (both financial and technological) which mask environmental degradation.[10] These characteristics have been widely recognized as critical challenges to the sustainability of industrial food systems.[11] Such challenges cannot be effectively addressed by the incremental changes associated with resistance or resilience adaptation strategies, but instead require a change in structure and function, a transformation, to a more sustainable and resilient US food system.[12]

This book shares the stories of twenty-five award-winning sustainable farmers and ranchers who offer us models of a resilient agriculture and glimpses of a resilient food future. These producers, like all producers everywhere, operate within a context shaped by cultural, social, environmental, technological and economic conditions. They are innovators of agroecosystems with the response capacity to remain productive and profitable largely without the support of and despite the barriers created by the growth and development of the agricultural industrial complex over the last fifty years.

These sustainable and resilient producers do more than simply grow healthy, high-quality food. Most have taken on all of the functions of the food system. Many manage the production of inputs: Nash Huber produces some of his own seed and compost; Brendon Rockey, Gail Fuller, Elizabeth Henderson and many others produce the nitrogen that feed their crops; Paul Muller, Ron Rosmann and Ken Dawson and many others produce natural pest suppression through management of biodiversity. Others process products on farm: Dan Shepherd cracks and cleans pecans; Will Harris and Greg Gunthorp slaughter, process and package livestock and make custom meat products; Tom Trantham processes and bottles milk products. Nearly all of these producers distribute and market directly to consumers.

These farmers and ranchers have worked to transform their local food systems without subsidies, promotional programs and access to production insurance; with limited support from research and technology development programs; and without subsidized marketing programs, tax breaks or waivers of environmental regulations. Some of these producers have found opportunity in state and federal agricultural and community development programs, while others are proud to say that they have never taken direct government payments of any kind.

All of these farmers and ranchers have enjoyed the support of a small but active community of sustainable agriculture producers, consumers, scientists, business people, technical advisors, educators, activists and others working together to create a more sustainable food system. With the help of this community, the producers featured in this book and many others like them all across our nation, have championed a new American food system—a sustainable food system that supports a triple bottom line of ecological, social and economic well-being on the farm and in the community. These farmers and ranchers have proven resilient while producing multiple benefits to the places they call home.

They have been successful despite an operating context that has placed formidable barriers in their path over the last fifty years. The systematic dismantling of regional processing and distribution infrastructure associated with the concentration and consolidation of the US food system has been and remains a barrier to meeting market demand for sustainable food. The focus of the US agricultural research, education and extension system on the problems of industrial food has limited public investment in research and development to address the needs of sustainable agriculture and food systems. Most recently, new barriers to sustainable food are being put into place by an increasingly onerous "one-size-fits-all" regulatory environment that seems unable to acknowledge the connection between biodiversity and human health, the broad community benefits of sustainable food systems or the crucial role that scale and production practices play in creating the environmental, social and economic harms of industrial food. What will it take to move the US food system onto a sustainable and resilient path? Simply put, we must transform the operating context of American agriculture. We can begin by building on the knowledge accumulated over more than 30 years of research and practice in sustainable food systems.

Understanding the Food System

In the latter half of the 20th century, a growing awareness of the natural resource degradation, animal welfare concerns, farm profitability challenges, farmworkers rights and food security issues drove a search for solutions that emerged as sustainable agriculture. Defined by Congress in 1990 and supported by a new federal program—the Sustainable Agriculture Research and Education program—much of the early investment in sustainable agriculture was focused on collaborative, on-farm research and development to improve agricultural production systems.

As the sustainable agriculture movement gained momentum through the 1980s and 90s, an increasingly global, concentrated and corporate US food system presented formidable barriers to the widespread adoption of sustainable production systems, and local food emerged as a sustainable solution.

From Land to Mouth

Brewster Kneen brought a Christian perspective to his 1995 analysis of the North American food system.[13] As he explored the principles of organization and behavior of the system, Kneen discovered an industrial logic that directly challenged food system sustainability. This logic was centered on three characteristics: distancing, uniformity and continuous flow. *Distancing* is the separation between producer and consumer created by processing, distribution and retailing. The globalization of the food system since the 1980s has dramatically increased distancing in the food system, making it difficult for consumers to understand the social or environmental impacts of their food choices. *Uniformity* is the focus of the industrial food system on processing cheap raw commodities into standardized foods. Uniformity reduces the freedom of farmers to select crops and production practices and leaves consumers with few meaningful choices at the supermarket. *Continuous flow* is the tendency for industrial food to flow from producer through the food system to the consumer, with few options for recycling or reuse of materials along the way. Continuous flow creates a lot of waste. According to Kneen, this "industrial logic," driven by distancing, uniformity and continuous flow, has eroded the ability of farmers and consumers to make choices consistent with Christian responsibilities to be good stewards of the land and to contribute to community well-being.

Kneen's solution was to reverse the logic of the industrial food system. Instead of distancing, promote proximity—shorten the distance, both literally and figuratively, between farmer and consumer so that food system impacts to land and community become community concerns; however, he warns that proximity as an organizing principle of the US food system would radically transform not just the food system but also settlement patterns, and would likely lead to decentralized human settlements—a re-regionalizing of both agriculture and population.

Kneen envisions a bioregion-based sustainable food system. He is primarily concerned with promoting a self-reliant food economy—a food system focused on feeding the people of the region with resources from that region in ways that promote the ecological, social and economic health of the region over time. He argues that a regional, self-reliant food system offers a comprehensive opportunity to deliver healthy food, air and water while reducing health care costs. Kneen advocates for healthy food as a basic human right and imagines government-supported food pharmacies, where one can go to fill prescriptions for healthy whole foods grown on local farms. He envisions agricultural land held in the public trust the way that our national wild lands have been protected in national and state parks.

Although Kneen does not mention climate change, resource scarcity or resilience in his analysis, his vision of a self-reliant, diverse and equitable food system organized within a bioregional framework meets all the criteria of a resilient agriculture. His work inspired new thinking about the structural relationships in a sustainable food system, taken up next by Kloppenburg and colleagues, who applied the idea of the foodshed to further develop Kneen's vision.

Coming into the Foodshed: The Commensal Community

In an article published in 1996, Jack Kloppenburg and his colleagues at the Michigan Center for Regional Food Systems argue that food is just one specific case of the general failure of "late capitalism" or post-industrialism to sustain the ecological, economic and social health of our communities. They point out that the failures of industrial food mirror similar failures in health, politics, finances and labor and argue that, ultimately, industrialism has driven an erosion of community. They go on to suggest that food, because it holds such a central place in our lives, is a good starting place for the restoration of community.[14]

In a call for the transformation of the industrial food system, Kloppenburg et al. applied the concept of the *foodshed*, the path that food takes from producer to consumer, to a sustainability analysis of the US food system.[15] Like Kneen, they find promise in a regionally focused, self-reliant food system. They expressed Kneen's ideas of proximity, diversity and balance through five principles that they argued are necessary for food system sustainability: moral economy, commensal community, self-protection, proximity and nature as measure.

Moral economy expresses the common expectations of mutuality, reciprocity and equity that guide simple human interactions in many cultures. This principle encourages economic exchanges that are conditioned by friendship, loyalty and pleasure in addition to cost and quality. Kloppenburg et al. argue that food is so central to the human expression of relationship that trading, preparing and sharing it should be structured to promote non-market social and cultural values as well as economic well-being.

Commensal community is the expression of the moral economy in a community. In ecology, commensalism describes a relationship between two organisms in which one eats without harming the other. In human terms, commensalism means building mutually supportive social relationships between people, as well as between people and the land and animals that feed us. Commensal community requires that other species and the land are treated with respect and regard for their well-being. This concept extends moral economy to the ecosystem within which the food system is rooted. This is akin to the "ethical stretch" that Wes Jackson suggested was required for a sustainable agriculture[16] and an expression of the land ethic, put in food system terms, that Aldo Leopold argued for so long ago.[17]

Self-protection is the principle that recognizes the challenges of working to create a commensal community within the existing industrial food system. Rather than fighting the system directly, Kloppenburg et al. envisioned transforming it from within by a reorganization of its social and productive capacity and the conscious transfer of resources from old relationships to new ones. They argue that food systems offer a diversity of opportunities for adjusting relationships as individuals as well as through businesses and organizations.

Much like Kneen, Kloppenburg et al. view *proximity* as a necessary condition for building meaningful relationships in which social and ecological welfare are immediate, practical and collective concerns rooted in a particular place. The last principle, *nature as measure,* encourages food system relationships that promote commensal relations with the land and animals that feed us. Natural resources define the place-based potential and the novel capacity for the emergence of commensal community. Together, the principles of *proximity* and *nature as measure* ask us to consider the question, "How do we feed ourselves in this place without harm?"

A Life Cycle Analysis of the US Food System

In 1999, the Center for Sustainable Systems at the University of Michigan facilitated the development of a set of indicators to evaluate the sustainability performance of the US food system. Aiming to create the space for a comprehensive and interdisciplinary discussion of agricultural sustainability, the Center invited representatives that reflected the diversity of agricultural perspectives typical of the Great Lakes/North Central Region.[18] Workshop participants included organic and conventional farmers, agribusiness, government, NGOs and agricultural scientists with expertise in sustainable agriculture and agroecology. Working together in a collaborative process, this diverse group identified a set of thirty-six indicators that could be used as a framework for a whole system sustainability assessment.

Center staff Martin Heller and Greg Keoleian used these indicators to evaluate the sustainability performance of the US food system.[19] This analysis viewed the US food system as an agroecosystem, with a boundary roughly equivalent to our national boundary. This approach encouraged the exploration of the food system as a coherent whole, an enormous agroecosystem that could be characterized by flow of energy and materials into, through and out of the national boundary. Taking this ecosystem perspective also put a focus on the relationships that make up the structure and function of the food system.

Heller and Keoleian's analysis raised some very basic questions. What is the purpose of the US food system? How does energy flow through the system? What kinds of materials does the system produce, where do those materials come from, and where do they go? Using the data gathered in their analysis, Heller and Keoleian were able to ask a fundamental question: Is the US food system sustainable?

Heller and Keoleian identified multiple sustainability concerns throughout the food system—from inputs to waste disposal—that led them to conclude that the US food system was not economically, socially or environmentally sustainable. Some key indicators supporting this conclusion include the loss of farmland to development, genetic diversity of crops and livestock, rates of soil loss and groundwater withdrawal, the age distribution of farmers, the low profitability of farm businesses, the legal and economic status of farm and food system workers, the rising costs of diet-related illness, the amount of edible food wasted and the fossil fuel intensity of the system.

While recognizing the preliminary nature of their analysis, Heller and Keoleian concluded that changing consumption behavior offers the most potential to enhance the sustainability of the US food system. They offer three specific recommendations to consumers: stop wasting food; eat fresh, unprocessed, locally sourced foods; and eat less meat and choose pasture-raised livestock products. They went on to explain that simply reducing food waste at the consumer level would reduce demand by nearly 30 percent, reverberating back through the food system in the form of less food retailing, distribution, processing and production and resulting in significant reductions in the associated environmental and social burdens of the system. They argue that consumer support for regional food systems would reduce the social and economic harms of corporate concentration, reduce dependence on fossil fuels and improve public health, while reducing the demand for meat and supporting pasture-based livestock production would result in significant improvements in the environmental and social impacts of industrial livestock production. While many of these recommendations seem commonplace today, remember that this study was conducted nearly twenty years ago.

Heller and Keoleian also noted the many social, economic and environmental impacts associated with the increasing consolidation and vertical integration throughout the input, production, processing and retailing phases of the food system in the latter half of the twentieth century. These trends have continued in the two decades since the study was conducted and continue to represent significant sustainability challenges particularly relevant today as we consider agricultural resilience to climate change and resource scarcity.

10.1. Industrial Concentration Erodes Food System Resilience

Because of its profound effects on the health and well-being of the nation, industrial concentration in US agricultural production, food processing and retailing has received increasing scrutiny by agricultural researchers, policy-makers and non-profit advocates over the last two decades. *Consolidation* is concentration within one sector of the food system, for example, when one seed company buys another seed company, leaving only one company where previously there had been two independent, competing businesses. *Vertical integration* is concentration across different sectors of the food system, for example, when one company purchases businesses that produce and sell inputs needed for livestock production, as well as meat processing and distribution businesses. This trend toward the concentration of resources suggests that the US industrial food system may be in the conservation phase of the adaptive cycle, suggesting that regeneration and renewal of the system will require a release of resources.

Consolidation has been a long-standing issue among livestock producers concerned about the lack of competition in the livestock industry. These concerns were first addressed in the 1890 Sherman Antitrust Act, in response to the growth of corporate monopolies in oil and steel, as well as agricultural goods like sugar, whiskey, beef and tobacco.[20] As the country industrialized and urban populations increased in the early part of the twentieth century, continued concerns about anticompetitive actions by meat packers led to the passage of the Packers and Stockyards Act in 1921.[21] Despite major federal efforts in the 1960s and again in the 2000s, today five multinational companies control more than 85 percent of the US meat supply.

Consolidation can lead to less competition and innovation in an industry, although it sometimes can increase sharing of knowledge between companies that were previously in competition.[22] As the diversity of businesses operating within a sector declines, there is less need for innovations to enhance marketability or reduce costs. Likewise, as the number of relationships between producers and other sectors of the food system declines, the choices of producers and food system workers are increasingly constrained by the demands of a limited number of buyers and employers.

Vertical integration, or cross-sector concentration, has also played a role in the declining fortunes of US agricultural producers and food system workers. A small number of multinational corporations have come to dominate the food system through informal alliances and ownership of businesses that link multiple sectors of the system. Vertical integration in the United States started in the poultry industry in the 1960s and began spreading rapidly to other food products in the 1980s. Although its costs and benefits are contested among economists, other social scientists examining questions of food system sustainability have discovered associated social, economic and environmental harms, particularly in industries that produce livestock in concentrated animal feeding operations, or CAFOs.[23]

Since the 1980s, consolidation and globalization has increased markedly in all phases of the US food system.[24] A wave of new consolidation since 2000 has led to oligopolies—markets in which the top four firms control more than 50 percent of market share—in the US seed, chemical, animal breeding and farm machinery industries,[25] grain industry,[26] food processing[27] and grocery retailing.[28] Much of this consolidation and vertical integration has been driven by the desire for greater energy and labor efficiency through economies of scale and scope, as well an effort to increase revenue in an industry plagued by an overproduction problem that has limited price increases. While consumers sometimes benefit from lower food costs, the limited research available has consistently shown that the savings realized from concentration are rarely passed on to the consumer.[29] In fact, consolidation in the grocery industry is usually associated with increased costs, reduced consumer choice and the emergence of food deserts in some communities.[30]

Creating Space for Sustainable Food

In 2002, Gail Feenstra shared lessons learned in more than a decade of experience participating in research and development of community food systems in California. Community food systems are defined as "a collaborative effort to build more locally based, self-reliant food economies in which sustainable food production, processing, distribution and consumption is integrated to enhance the economic, environmental and social health of a particular place."[31] Akin to Kloppenburg et al.'s principle of self-protection, Feenstra identified four spaces that were consistently associated with successful community food system projects: social space, political space, economic space and intellectual space.

Social space was crucial for the cultivation of the social capital that fueled the success of many food system projects. This might be physical space, like a farmers market or a community garden, but more often was just creating the time and space in which to make plans, discuss community concerns, develop a common vision and celebrate project successes. Equally critical to a project's success is the cultivation of *political space.* Working with community leaders and policy-makers helps to build the relationships needed to institutionalize food system changes. Community organizing is key to the cultivation of political space. *Intellectual space* is created through the cultivation of interdisciplinary partnerships and town/gown collaborations that facilitate social learning. *Economic space* is created through financial support for food system projects during planning, initial startup and development.

In the first decade of this century, sustainable food system projects focused on expanding community-based capacity for food production, processing and marketing, while linking producers, processors, retailers and consumers at local and regional levels. Physicians, dieticians, public health specialists and municipal planners joined the sustainable food movement during this time, bringing more focus to land use, transportation and economic development issues and promoting increased availability and access to healthy, nutrient-dense foods.[32] By mid-decade, civic agriculture emerged as a new conceptual framework for community-based food systems.[33] This framework extended the concept of sustainable agriculture to emphasize the potential of agriculture and food systems to produce multiple benefits in support of the social well-being of communities—particularly the capacity of local food production to restore local economies and revitalize democratic participation.

The Rise of Local Food

Local food became a concern of the general public in the mid-2000s with the publication of two popular books: Michael Pollan's *The Omnivore's Dilemma* and Barbara Kingsolver's *Animal, Vegetable, Miracle.* Pollan explored the impacts of overproduction on our food system and our health, pondered the oxymoron of "industrial organic" and shared compelling models of sustainable agriculture. Kingsolver's work shared the personal experience of her family's year of eating locally. Both authors drew on a large body of food systems scholarship that seemed to be pointing to local food—food produced within fifty miles, a hundred miles, four hundred miles of home—as a condition of sustainability. A new word—*locavore*—was added to the Oxford American Dictionary in 2007, seasonal eating and regional cuisine became a new trend, and millions of consumers connected more directly with the people growing their food and local food traditions, many for the first time.

The spectacular entrance of the Transition Initiative[34] into community-based redesign for resilience introduced a new perspective on food system challenges. The Transition Initiative champions an innovative model of community-based action to motivate holistic solutions to the interlinked challenges of climate change and our addiction to fossil fuel. Originator Rob Hopkins synthesized best practices in addiction recovery, resilience theory and permaculture design to create community-based processes for sustainable and resilient community redesign. Food systems have been a focus for many Transition Initiative projects, including those underway throughout the United States.[35]

As food system research expanded and experience from local projects began to accumulate, some of the sustainability benefits assumed by the proponents of local food systems came into question.[36] Limited research into energy use found that local food systems may be more energy-intensive than food sourced regionally and found some cases in which food sourced internationally had a lower energy input than local food. As food miles research developed, it became clear that although food sourced locally traveled fewer miles, the environmental impacts of food transport are complex and food miles alone is not a useful indicator of sustainability.[37]

Others became concerned about an erosion of values as the local food movement gained momentum. Some wondered if local food was headed down the same path as "industrial organic"[38] when Walmart, the largest grocer in the United States, announced it would begin selling local food. The introduction of food hubs and other means to increase the volume of local food moving through regional markets raised concerns that local food was being sacrificed at the altar of cheap food and motivated reminders that local food is defined by more than just geography.[39] Local food produces many social benefits that may be lost as the distance between the producer and consumer increases. For example, new research suggests that the climate resilience benefits often associated with local food systems may not survive scaling up because distancing erodes the social capital cultivated by direct markets.[40]

As scaling up became a focus of the local food movement, a large group of full-time, mid-scale family farmers left behind by the global industrial food system began to claim the attention of local food advocates. This disappearing sector of mid-scale farms and related agri-food enterprises—known as the Agriculture of the Middle (AOTM)[41]—produce at volumes too small to market as a bulk agricultural commodity, but too large to market directly to consumers. The most successful AOTM farms participate in cooperative business alliances[42] that distribute to regional markets. These alliances add value to their products by differentiating them in some way—perhaps by production region or a certification program. Proponents of cooperative alliances argue for values-based food supply chains[43] as a way to preserve the social, economic and ecological benefits associated with local foods while serving regional markets. Demand for these distinctive products from regional supermarkets, restaurants, public and private institutional buyers and individual consumers is strong and continues to grow.

Remaking the North American Food System: Challenge and Opportunity

The 2007 publication *Remaking the North American Food System* describes the progress, challenges and opportunities for sustainable agrifood systems.[44] In a concluding discussion,[45] Clare Hinrichs and Elizabeth Barnham report that two decades of community-based research and development efforts have enjoyed some success, but acknowledge that efforts to transform the industrial food system have been hampered by confusion about core sustainability concepts and difficulty securing support for the systems research needed to discover holistic solutions to the problems of industrial food. They point to national organizations like the Farmers Direct Marketing Association, state and regional direct marketing associations, food policy councils and new federal marketing programs like *Know Your Farmer, Know Your Food* as fruitful ground for the further development of local food systems but caution that this trend toward a singular focus on economic relationships limits the potential to realize the multiple benefits of sustainable food. They identify a number of other trends that may accelerate the transition to sustainable food systems, such as a shift in political support for agricultural subsidies, new concerns about the public health harms of industrial foods, increasing concerns about climate change, the vulnerability of the global food system to terrorism and the growing scarcity of natural resources such as energy, land, water and ecosystem services.

The Sunfood Agenda: A Comprehensive Solution to 21st Century Challenges

In an open letter to President Obama published in the *New York Times* just after the election in 2008, Michael Pollan proposed sustainable regional food systems as a comprehensive solution to the triple challenge of peak oil, climate change and public health facing our nation.[46] Pollan identified the sustainability challenges of food systems in the United States and abroad as key public health and national security concerns. He warned that little progress would be possible on the nation's public health, energy independence or climate change challenges without placing a priority on addressing the harms of industrial food. His "Sunfood Agenda" called for federal policy changes to resolarize agriculture, re-regionalize food systems and rebuild America's food culture.

To achieve the resolarization of American agriculture, Pollan focused on federal programs to transition to diversified integrated production systems that conserve biodiversity and enhance natural resource quality on farms. He recommended support for the development of perennial grains, pasture-based livestock production, training for new farmers, farmland conservation and the development of effective systems for recycling of food and other organic wastes back to farmland. Pollan also recommended an end to all direct and indirect federal support for confined animal feeding operations (CAFOs).

Pollan recognized that the rest of the food system—the operating context—would have to be reorganized to support a resilient and sustainable American agriculture and presented proposals to support the re-regionalization of US agriculture. Some specific proposals include the development of food safety regulations sensitive to scale and marketplace; the creation of a meat inspector corps to support portable slaughter and processing equipment adapted to small-scale livestock production; regionalizing federal food procurement; rebuilding regional processing, storage and transport networks; and creating a federal definition of "food" based on nutrient density.

Recognizing the crucial role consumer support would play in achieving the transformation to a re-solarized and re-regionalized food system, Pollan proposes to cultivate a new American food culture. He recommends initiatives to increase food literacy such as programs that support urban agriculture; the growing, preparation and consumption of whole foods in public schools and offering product life cycle impacts on food packaging. He also recommends shifting responsibility for federal nutritional advice from the USDA to the Surgeon General and encouraging the First Family to model the new American food culture through the use of seasonal menus, meatless meals and the establishment of an urban farm on the White House lawn.

Pollan's proposals draw on more than 30 years of intellectual inquiry, collaborative research and practical experience devoted to understanding sustainable food systems, and many of his proposals build upon existing local and regional models. Although there are some critical gaps in the agenda, particularly the lack of attention to food security issues and agricultural justice for farmers and food workers, the Sunfood Agenda offers us a credible vision of a sustainable and resilient food future and some practical steps to get us there.

Adding Resilience to the Menu

In the years since the Sunfood Agenda was proposed, a new awareness of climate change, coupled with the global financial and oil shocks, brought additional urgency to the quest for sustainable solutions. New federal programs, private donors and NGOs, research programs and publications in scientific journals and the popular press began to address agricultural and food system adaptation to climate change and other twenty-first-century resource challenges using the language of resilience science.[47] Major US cities and intrastate regions, several states and a number of multistate projects have worked to understand food system sustainability, assess the productive capacity of their food systems and develop plans to promote more sustainable and resilient food systems.[48] As experience and knowledge at different scales—from the local to global—began to accumulate, the regional scale emerged as the right size for a sustainable and resilient food system.[49]

Regional Planning for Resilience

Regions have a number of qualities that, taken together, offer a unique opportunity for addressing food system sustainability and resilience.[50] Our natural understanding of the physical world is based on biophysical patterns defined at regional scales, for example ecosystems and watersheds. Strong cultural dimensions often arise at a regional scale, and useful units for governance are often found there as well. A regional focus will typically include both urban and rural areas, offering an opportunity to include intra-regional interactions such as trade, development, population, transportation networks and other system elements that are likely to become more important as urban areas increase in significance in coming years. Finally, regional populations have influence at the national scale and through democratic participation have a voice in setting national policies and programs that can be used to create the operating conditions needed to promote a sustainable and resilient US food system.

Noteworthy regional projects exploring food system resilience in North America include those based in Iowa, Vermont and New England. These plans are among the first to articulate resilience as an objective and to develop the resilience criteria needed to support adaptive management of a sustainable food system.

The Iowa Food Systems Council (IFSC) engaged food system stakeholders across the state to develop a comprehensive approach for assessing food system resilience.[51] Released in 2011, this innovative plan presents a sustainability and resilience assessment of the Iowa food system and recommendations for policy, program and research strategies that would improve the performance of the system on sustainability and resilience goals. The IFSC is currently preparing their first report card on the performance of the Iowa food system on the plan's indicators, though this work has been hampered by a lack of resources to complete regular updates as initially planned.[52]

Also in 2011, Vermont's Center for Agricultural Economy released a comprehensive strategic plan designed to enhance economic development in the Northeast Kingdom region of the state through the growth of local food systems.[53] The planning process was based on a soil-to-soil, closed-loop food system model and is consistent with the goals of Vermont's Farm to Plate Investment Program.[54] The plan presents a practical and credible path to achieve broad sustainability and resilience goals for the Northeast Kingdom food system. Many of the sixty targets (indicators) for assessing progress toward the plan's ten core goals enhance food system resilience: local provision of production inputs and recycling of farm and food wastes; support for diversified and profitable local production; development of infrastructure sufficient to supply year-round consumer demand for some food items; food system processes that serve to enhance environmental quality; agriculture and food system skills development; and local provision of healthy, fresh, affordable local food for all residents.

10.2. The "Soil to Soil" Food System Model[55]

The "Soil to Soil" food system model is noteworthy because it extends the concept of the food system to include waste management. In this model, organic farm and food wastes are composted and returned to farms to improve soil quality or used to produce bioenergy. This is an important improvement in food system management, but still leaves out a large source of nutrient loss: human manure. Practical approaches to managing human wastes as part of the food system waste stream are controversial because of safety concerns and strong cultural barriers, but offer the potential for multiple economic, social and environmental benefits.

The urine-diverting dry toilet (UDDT) is a new approach to the problem getting a lot of attention around the world.[56] Without using any water, UDDTs separate urine from feces to facilitate recycling of both materials back to farmland. Vermont's Rich Earth Institute is conducting the nation's first formal experiment in community-scale nutrient reclamation using source-separated urine to fertilize hay fields near Brattleboro.[57]

The New England Food Vision 2013 presents a detailed asset analysis of a sustainable and resilient regional food system in 2060. The vision aligns with four core values that place an emphasis on food security and sustainability: everyone has access to adequate food, everyone enjoys a

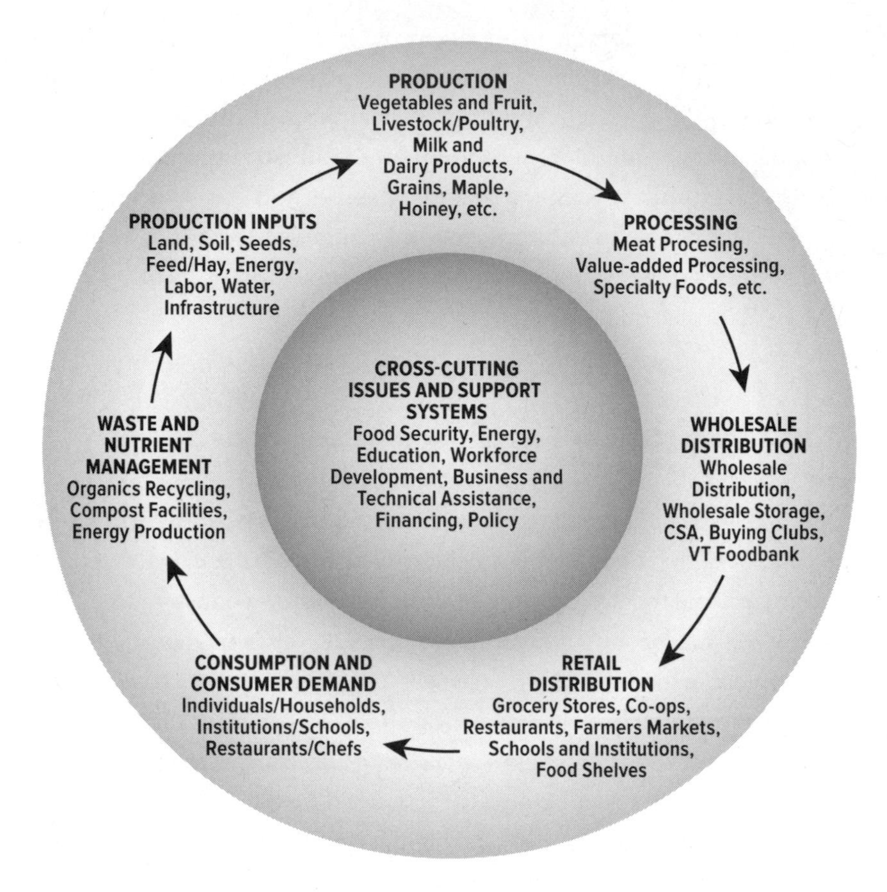

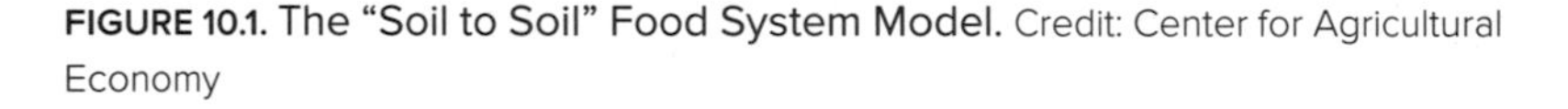
FIGURE 10.1. The "Soil to Soil" Food System Model. Credit: Center for Agricultural Economy

healthy diet, food is sustainably produced, and food helps build thriving communities. The New England Food Vision explores sustainable food production, including natural resource concerns relevant to energy and water resources, climate change and biodiversity and social concerns such as building food-system capacity and farm and food industries. Although it does not present a plan or performance metrics, the Vision does reference Vermont's Farm to Plate strategic plan. A companion report[58] analyzes policy barriers and gaps associated with the Vision's goal of increasing production and consumption of New England-sourced food and identifies policy changes to support expanding production, strengthening food supply chains and enhancing multistate cooperation toward a more robust and resilient regional food system.

This recent work in regional food system planning offers a glimpse of a commensal food future by proposing models of sustainable and resilient food systems in answer to the question,"How do we feed ourselves in this place without harm?"

Cultivating Food Systems for a Changing Climate

What is the vulnerability of the US industrial food system to climate change? Thinking about this question raises more questions than answers: What are the main climate-related threats to North American food systems, both now and in the near future? What are the food system sensitivies to these threats? What is the adaptive capacity of our food system? How can our current understanding of food system sustainability and resilience contribute to effective climate change adaptation options?

While we are only now beginning to address these questions as a nation, our understanding of sustainability and resilience suggests that the US industrial food system does not have ecological or social response capacity to persist in a changing climate. A transformation of the system will likely be required to sustain the US food system into the 21st century. Can we find opportunity in the climate change challenge to cultivate a new kind of US food system? A food system that produces abundant, nutrient-dense food while restoring healthy natural resources and rebuilding communities? A food system that helps to mitigate global warming and protect us from weather variability and extremes while remaining productive in a changing climate?[59] What would such a food system look like?

The concept of the agroecosystem[60] can aid our visioning of a resilient US food system. Recall that, like the ecosystem, the scale and boundary of the agroecosystem is flexible and defined by the user. We can expand the concept of the agroecosystem beyond the individual farm to include the farm community, the region or even the nation and employ resilience criteria[61] as a framework for regional food system assessment and design. Just as sustainable agriculture principles helped us to envision a resilient agriculture, the principles of sustainable food systems help us to envision a resilient US food system.

A resilient and sustainable US food system will rest on a web of food systems embedded in a particular place, a civic agriculture defined perhaps by bioregion,[62] watershed or metropolitan area. This New American Food System will support diverse food supply webs oriented to local and regional markets.[63] Resilience emerges through the diversity of linkages within and across multiple scales, from the local to the global, and the application of regenerative ecological principles and adaptive management strategies to food system design and development. Agricultural inputs, production, processing, distribution and sales are focused on regional markets, and food trade is limited to excess production and to a few specific products that are well adapted to the region.

Taking a cue from nature, these food webs will produce their own energy through sustainable energy systems and recycle wastes to close regional nutrient cycles. This food web will be less labor efficient and may be less land efficient, but it will be vastly more energy- and water-efficient. New measures of food system performance will value the production of fresh, nutrient-dense foods and the cultivation of ecological and social response capacity. This resilient food web will produce multiple environmental, social and economic benefits at the local and regional scale. Although many questions remain about how this system might evolve over time, we know enough, right now, to begin to cultivate a sustainable and resilient food system for a changing climate.

10.3. A Nationally Integrated Regional Food System

One proposal for the New American Food System envisions a redirection of resources toward the development of a nationally integrated regional food system.[64] Large metropolitan areas provide natural nodes for food systems focused on supplying regional food demands (see Figure 10.2).

Metropolitan areas present a unique operating environment with many benefits and some challenges to a sustainable and resilient agriculture.[65] On

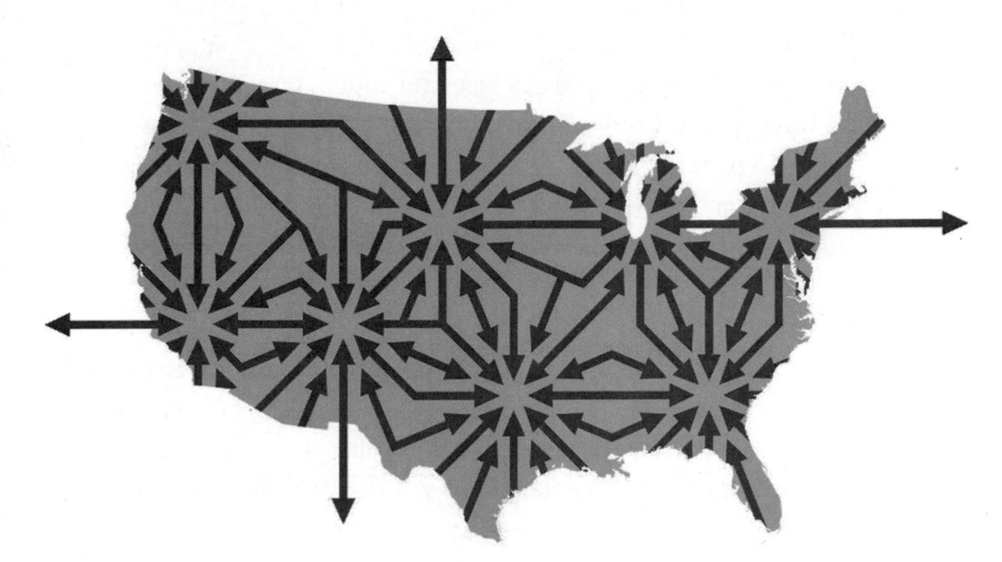

FIGURE 10.2. A Nationally Integrated Regional Food System Model. Credit: Urban Design Lab, Reprinted with permission.

the plus side, sustainable producers located in or near these areas have easy access to a large population of potential customers, high-value direct markets and value-added processing opportunities. Physical infrastructure for power, water, transportation and other resources is usually well-developed, and proximity to the metropolitan core offers opportunities for meaningful off-farm employment for non-farming family members. Metropolitan farming also presents some challenges. Land values are high, non-farming residents may object to farming operations, and access to traditional farm services may be difficult.

About 39 percent of all US farms are located in metropolitan areas; these farms account for about 40 percent of the value of US agricultural production. Metro farms have a different product mix than farms in non-metro areas; high-value crops and dairy products make up a larger share of their production, while cash grains and beef make up a smaller share. The top ten metro areas in order of population size:[66] New York, Los Angeles, Chicago, Dallas, Houston, Philadelphia, Washington DC, Miami, Atlanta and Boston.

The Way Forward

Resilience practice suggests that successful system transitions are generated through leadership at the community-scale for a diversity of innovations encouraged by government through policy changes that remove barriers and provide incentives. This community-based approach promotes the diversity, modularity and accumulation of capital assets that enhances the resilience of the system.[67]

A useful framework to guide the transition of the US food system may be one that describes a continuum of sustainability and resilience qualities associated with food systems (see Figure 10.3). The sustainability continuum describes the capacity of the food system to produce multiple benefits across all three dimensions of sustainability—ecological, social and economic. The resilience continuum describes the capacity of the food system to produce the diversity, modularity, tight feedbacks and capital assets associated with response capacity.

Industrial food systems can be characterized as focused primarily on economic growth and the production, processing and distribution of highly processed commodity products for national and global networks. These systems, like most industrial systems, operate under the assumptions of a stable climate and the unlimited global flow of energy, and are managed for optimum conditions. Materials and technology typically substitute for ecosystem services in the production, processing, distribution and waste disposal functions. Some of the practices that shape the identity of industrial food systems include the use of labor- and land-efficient technologies, genetically modified crops, synthetic fertilizers and pesticides; geographic concentration and specialization in production, processing and distribution; vertical integration; and globally sourced, carbohydrate-dense, highly processed foods.

Sustainable food systems seek to achieve environmental, social and economic goals by producing, processing and distributing value-added products to local and regional markets. Like industrial systems, they also operate under the assumption of stable climate and unlimited global flow of energy and materials and seek optimum system configurations; however, ecosystem services typically substitute for the energy and materials used in the production, processing, distribution and waste disposal functions of industrial food systems. Some of the practices that shape the identity of sustainable food systems include integrated and diversified farming systems, pasture-based livestock production, local and regional processing and consumption of fresh and seasonal foods.

Both industrial and sustainable food systems also lie along a resilience continuum. Industrial food system responses to the challenges of climate change are already underway and include strategies such as the development of climate-ready crops and climate-smart farming systems.[68] This approach to resilience relies on enhancing technological assets to manage climate disturbances and moves the food system horizontally along the resilience axis. Sustainable intensification is another adaptation strategy currently receiving a lot of attention. This approach cultivates technological capacity to reduce the environmental impacts of the industrial food system and moves the food system vertically along the sustainability axis.

While both of these approaches signal an acceptance within the industrial food system that the assumption of stability is no longer valid, they are examples of of maladaptation, because they involve investing in adaptation strategies that protect the existing food system from climate disturbances and resource shocks without cultivating ecological, human, or social capacity for resilience.[69] Such strategies may extend the life of the existing food system, but as the pace and intensity of climate change increases, they will become increasingly expensive, less effective and more likely to fail. The danger of staying on this path is that we may squander the resources needed to achieve a successful transition to a more desirable sustainable and resilient food system.

Sustainable resilient food systems are represented in the upper right quadrant of Figure 10.3. These systems are characterized by a focus on eco-social well-being, the production of value-added foods for local and regional markets, the recognition that stability can no longer be assumed and the search for robust food system organization. They emphasize regional self-reliance and the cultivation of ecological and human capacity

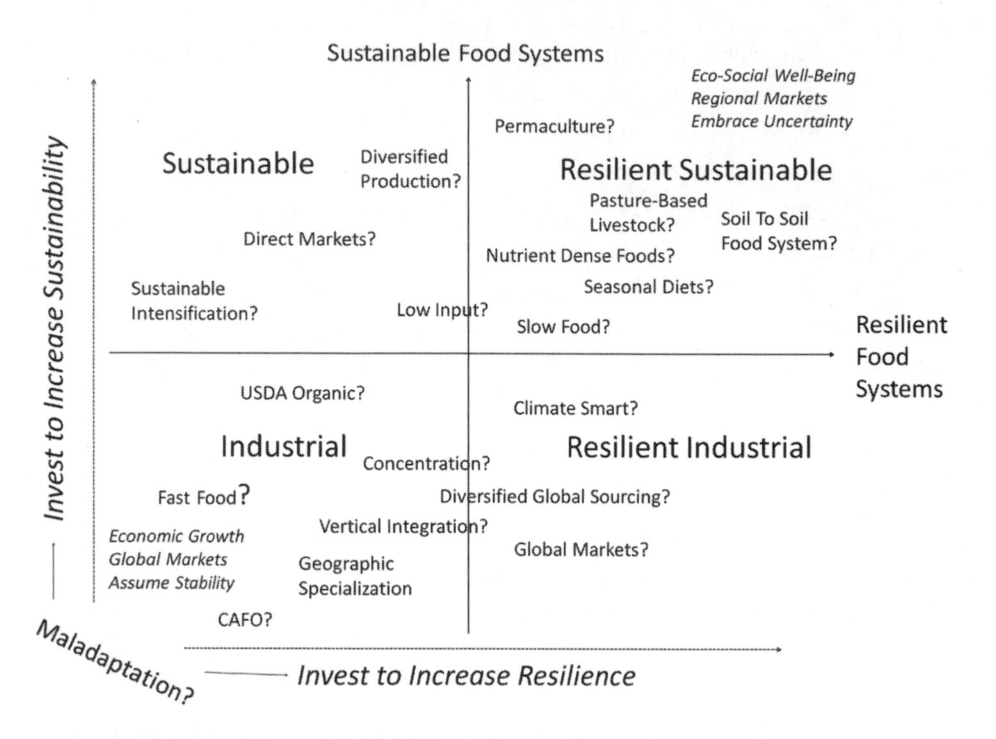

FIGURE 10.3. Where Are We? Where Do We Want to Go? Sustainability and resilience are two different but complementary qualities of agroecosystems that vary across a sustainability/resilience space. Food system models produce sustainability and resilience outcomes that can be located in this space. Investments that move food systems only on the horizontal or vertical plane are likely to be maladaptations, because effective adaptations to climate change cultivate both sustainability and resilience.

as an adaptation strategy. Resilience theory suggests that such systems can be sustained through time because they have high ecological and social response capacity and so can evolve with changing conditions.

The US industrial food system is not designed to produce the ecological and social response capacity required to support a sustainable and resilient food future under changing climate conditions. The sustainable and resilient way forward requires a transformation, a change in form and function and, perhaps most importantly, a change in purpose of the US food system to one of sustainability and resilience. The path to this system does not lie along a horizontal or a vertical plane, but rather along the diagonal, as shown in Figure 10.4. It is not one path but many

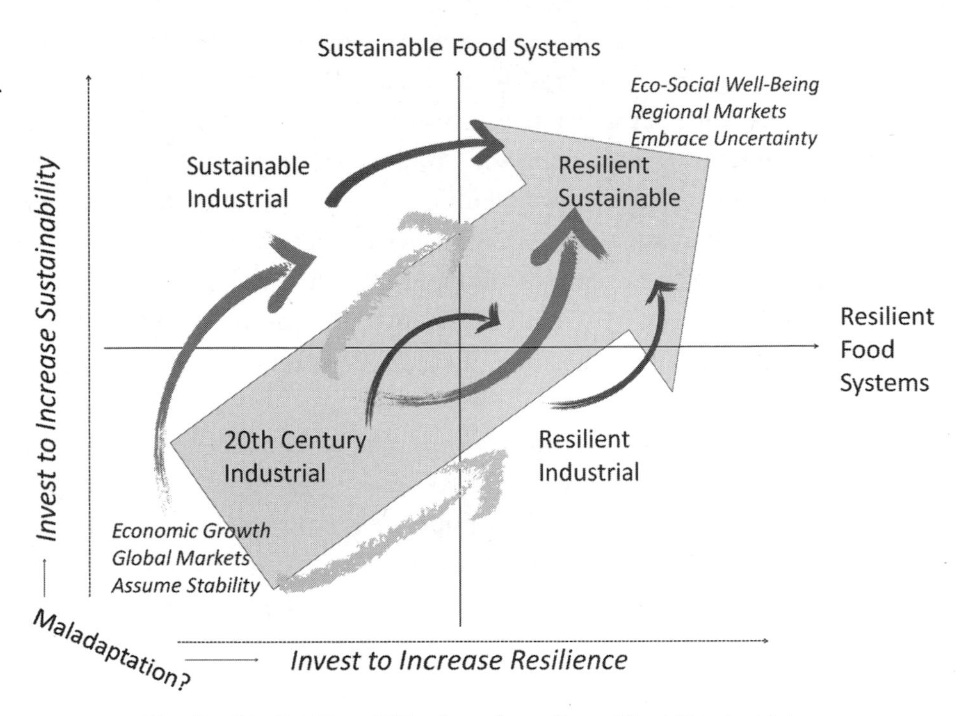

FIGURE 10.4. How Do We Get There? The transformation of the US industrial food system lies along a path that cultivates both sustainability and resilience achieved through a diversity of place-based adaptations that move us towards the goal of a nationally-integrated network of sustainable and resilient regional food systems based on a soil-to-soil model.

paths, not a silver bullet but silver buckshot,[70] a diversity of individual adjustments and adaptations of components within our food system that together move it from the current vulnerability to a more sustainable and resilient future. We have the knowledge and experience—the seeds of the new American food system—on the ground and growing in many parts of the country.

A first step to put us on a path toward a more sustainable and resilient US food system is to begin to change the operating context within which food and farm businesses make decisions.[71] The overall strategy is to begin redirecting investments away from those that simply protect the existing food system from disturbance. We must begin to invest in the development and accumulation of capital assets that enhance the sustainability and resilience of the US food system.

We can begin now to redirect private and public investment toward the cultivation of ecological, human, and social response capacity in our food system. We can begin now to work toward achieving the goals of the Sunfood Agenda, along with food security and agricultural justice, through regional planning that improves upon the innovative first steps made in Iowa, Vermont and New England. We can choose to view the climate change challenge as an opportunity to take a step back, reconsider the desirability of our current food acquisition strategy and then step forward onto a transformative path that takes us into a sustainable and resilient food future.

As we consider this path, it's important to remember that we are not going back to the future. Since colonial times, American agriculture has eroded community sustainability and resilience. Important also to understand that resilience is not just another name for sustainability. Resilience changes the rules of the game. It changes the way we define problems and the way we evaluate solutions.

Resilience requires us to see with new eyes to ask different kinds of questions, to embrace uncertainty and to find opportunity in change. It requires a fundamental shift in the way we organize our lives. We must accept that we cannot burn or build our way to a solution, that we cannot depend on human ingenuity alone, but finally acknowledge our deep dependence on the natural world. We are stepping forward into something new, a future that has the potential to be as different from our current reality as agriculture was to our foraging ancestors.

We can cultivate a new kind of food system, a food system that has the capacity to protect us from the inevitable challenges ahead, a food system capable of producing so much more for our communities than the industrial food system of today. We know enough to take the first steps along the path to a sustainable and resilient future. Let's begin.

Endnotes

1. W. Berry. 1990. The Pleasures of Eating, in *What Are People For?* North Point Press: New York.
2. M. Miller et al. 2013. Critical Research Needs for successful food systems

adaptation to climate change. *Journal of Agriculture, Food Systems and Community Development,* 3(4)161–175.

3. G. Berardi, R. Green, and B. Hammon. 2011. Stability, sustainability, and catastrophe: Applying resilience thinking to U. S. agriculture. *Human Ecology Review,* Vol. 18 (2): 115–125.
4. For example, see Oxfam. 2009. Exposed: Social vulnerability and climate change in the US Southeast. oxfamamerica.org/explore/research-publications/exposed-social-vulnerability-and-climate-change-in-the-us-southeast/
5. See chapter 2.
6. USDA Expert Stakeholder Workshop for USDA Technical Report on Global Climate Change, Food Security and the U.S. Food System. 2013. globalchange.gov/sites/globalchange/files/Climate%20Change%20and%20Food%20Security%20Expert%20Stakeholder%20Mtg%20Summary%20(Final).pdf
7. H. Eakin. 2010. What Is Vulnerable. In J. Ingram et al. (eds). *Food Security and Global Environmental Change.* Earthscan: Washington, DC.
8. J. Wong and R. Schuchard. 2011. Adapting to Climate Change: A Guide for Food, Beverage, and Agriculture Companies. Business for Social Responsibility: Industry Guides.
9. USDA Expert Stakeholder Workshop for USDA Technical Report on Global Climate Change, Food Security and the U.S. Food System. 2013. globalchange.gov/sites/globalchange/files/Climate%20Change%20and%20Food%20Security%20Expert%20Stakeholder%20Mtg%20Summary%20(Final).pdf
10. P. Ericksen. 2008. What Is the Vulnerability of a Food System to Global Environmental Change? *Ecology and Society,* 13 (2): 14.
11. National Research Council. A Pivotal Time in U.S. Agriculture. Ch. 2 in *Toward Sustainable Agricultural Systems in the 21st Century. 2010*. National Academies Press: Washington, DC.
12. For example, Reganold et al. 2011. Transforming U.S. Agriculture. *Science,* Vol (322): 670–671; D. Giovannucci. 2012. Food and Agriculture: The Future of Sustainability. Sustainable Development in the 21st Century. UN Dept. of Economic and Social Affairs, Div. for Sustainable Development.
13. B. Kneen. 1995. *From Land to Mouth: Understanding the Food System,* Second Edition. NC Press Limited.
14. J. Klopenburg et al. 1995. Coming Into the Foodshed. *Agriculture and Human Values,* 13: 33–42.
15. Modeled on the watershed concept, the concept of the foodshed offers a way to visualize the geographic area through which food flows from producers through the food system to consuming communities. Walter Hedden, Chief of the Commerce Bureau of the Port of New York Authority, is credited with the first use of the term in his 1929 book, *How Great Cities Are Fed,* a

comprehensive assessment of the New York City food supply. Hedden's analysis found that, by the 1920s, New York's fruit and vegetables were already travelling an average of 1,500 miles. More than 60 years after Hedden coined the term, the foodshed concept emerged again in discussions about the sustainability benefits of foods sourced locally and has remained a useful and popular frame for engaging in food system sustainability. C. Peters et al. 2009. Foodshed Analysis and Its Relevance to Sustainability. *Renewable Agriculture and Food Systems*, Vol. 24 (1):1–7.

16. "It seems that we have forever talked about land stewardship and the need for a land ethic, and all the while soil destruction continues, in many places at an excelerated pace. Is it possible that we simply lack enough stretch in our ethical potential to evolve a set of values capable of promoting a sustainable agriculture?" W. Jackson. 1980. A Failure of Stewardship. Ch. 1 in *New Roots for Agriculture*. New Edition, Bison Books.
17. "All ethics so far rest upon a single premise: that an individual is a member of a community of interdependent parts. His instincts prompt him to compete for his place in that community, but his ethics prompt him also to cooperate (perhaps in order that there may be a place to compete for). The land ethic simply enlarges the boundaries of the community to include soils, waters, plants and animals, or collectively: the land." A. Leopold. 1949. The Land Ethic, p. 171, in *A Sand County Almanac, With Essays on Conservation*. Oxford University Press: New York.
18. G. Aistars (ed.). 1999. A Life Cycle Approach to Sustainable Agriculture Indicators. Center For Sustainable Systems, Univ. of Michigan. umich.edu/~nppcpub/resources/compendia/Proceedings.PDF
19. M. C. Heller and G. A. Keoleian. 2003. Assessing the sustainability of the US food system: A life cycle perspective. *Agricultural Systems*, 76: 1007–1041.
20. The Sherman Antitrust Act. 2005. *West's Encyclopedia of American Law*. encyclopedia.com/topic/Sherman_Antitrust_Act.aspx
21. Packers and Stockyards Act. 2007. USDA. gipsa.usda.gov/Publications/psp/broch/psact.pdf
22. For example, K. Fuglie et al. 2012. Rising Concentration in Agricultural Input Industries Influences New Farm Technologies. Amber Waves, USDA-ERS.
23. National Research Council. A Pivotal Time in U.S. Agriculture. Ch. 2 in *Toward Sustainable Agricultural Systems in the 21st Century*. 2010. National Academies Press: Washington, DC.
24. T. Lyson. 2004. Going Global. Ch. 3 in *Civic Agriculture: Connecting Farm, Food and Community*. Tufts University: Boston.
25. K. Fuglie et al. 2012. Rising Concentration in Agricultural Input Industries Influences New Farm Technologies. Amber Waves, USDA-ERS.

26. Concentration in the Food Industry. 2014. CorpWatch. community.corpwatch.org/adm/pages/food_industry.php
27. Ibid.
28. Food and Water Watch. 2010. Consolidation and Buyer Power in the Grocery Industry Factsheet. documents.foodandwaterwatch.org/doc/RetailConcentration-web.pdf
29. Economic Concentration and Structural Change in the Food and Agriculture Sector: Trends, Consequences and Policy Options. 2004. Democratic Staff of the Committee on Agriculture, Nutrition, and Forestry, United States Senate.
30. Food and Water Watch. 2010. Consolidation and Buyer Power in the Grocery Industry Factsheet. documents.foodandwaterwatch.org/doc/RetailConcentration-web.pdf
31. G. Feenstra. 2002. Creating Space for Sustainable Food Systems: Lessons From the Field. *Agriculture and Human Values*, 19:99–106.
32. C. Koliba, E. Campbell, and H. Davis. 2011. Regional Food Systems Planning: A Case Study from Vermont's Northeast Kingdom. Food System Research Collaborative, Opportunities for Agriculture Working Paper Series, Vol. 2, No. 2. Center for Rural Studies, Univ. of Vermont.
33. T. Lyson. 2004. *Civic Agriculture: Reconnecting Farm, Food and Community*. Tufts University: Boston.
34. R. Hopkins. 2008. *The Transition Handbook: From Oil Dependency to Local Resilience*. Green Books: London.
35. Transition US is a non-profit organization that provides inspiration, encouragement, support, networking and training for Transition Initiatives across the United States. transitionus.org/home
36. C. Peters et al. 2009. Foodshed Analysis and Its Relevance to Sustainability. *Renewable Agriculture and Food Systems*, Vol. 24 (1): 1–7.
37. H. Hill. 2008. Food Miles Background and Marketing. ATTRA National Sustainable Agriculture Information Service.
38. J. Ikerd. 2008. Reclaiming the Heart and Soul of Organics. web.missouri.edu/ikerdj/papers/Boulder%20Organic%20Summit%20--%20Soul%20of%20Organic.htm
39. For example: B. Born and M. Purcell. 2006. Avoiding the Local Trap: Scale and Food Systems in Planning Research. *Journal of Planning Education and Research*, 26:195–207; J. Johnston et al. 2009. Lost in the Supermarket: The Corporate-Organic Foodscape and the Struggle for Food Democracy. *Antipode*, Vol. 41 (3) 509–532; L. Delind. 2010. Are local food and the local food movement taking us where we want to go? Or are we hitching our wagons to the wrong stars? *Agric Hum Values*, DOI 10.1007/s10460-010-9263-0;

A. Perrett. 2013. Beyond Efficiency: Reflections from the Field on the Future of the Local Food Movement. https://www.uvm.edu/foodsystems/summit/Perrett-BeyondEfficiency.pdf

40. C. Furman et al. 2013. Growing food, growing a movement: Climate adaptation and civic agriculture in the southeastern United States. *Agric Hum Values*. DOI 10.1007/s10460-013-9458-2.
41. F. Kirschenmann et al. 2005. Why Worry About the Agriculture of the Middle? A White Paper for the Agriculture of the Middle Project. agofthemiddle.org/archives/2005/08/why_worry_about.html.
42. A cooperative alliance is a self-organized association of persons united to meet their common economic, social and cultural needs and aspirations through jointly owned and democratically controlled enterprises. The U.S. is home to 30,000 cooperatives with more than 256 million members representing all sectors of the economy, including agriculture, banking, consumer, fisheries, health, housing, insurance and workers. International Cooperative Alliance, ica.coop/en/whats-co-op.
43. Food value-chains are strategic cooperative business alliances formed to distribute significant volumes of high-quality, differentiated food products produced by mid-scale farmers and ranchers. See, "Values-Based Food Supply Chain Case Studies," Center for Integrated Agricultural Systems. University of Wisconsin, Madison. cias.wisc.edu/aotm-case-studies/
44. C. Hinrichs and T. Lyson (ed). 2007. *Challenges and Opportunities in Remaking the Food System*. University of Nebraska Press: Lincoln.
45. C. Hinrichs and E. Barham. 2007. Conclusion: A Full Plate. In *Challenges and Opportunities in Remaking the Food System*. C. Hinrichs and T. Lyson (ed). University of Nebraska Press: Lincoln.
46. M. Pollan. 2008. Farmer in Chief. *New York Times Magazine,* Oct. 12. nytimes.com/2008/10/12/magazine/12policy-t.html?pagewanted=all
47. For example: C. King. 2008. Community Resilience and Contemporary Agri-Ecological Systems: Reconnecting People and Food, and People with People. *Systems Research and Behavioral Science Syst. Res,* 25, 111–124; Whole Measures for Community Food Systems: Values-Based Planning and Evaluation. 2009. Center for Whole Communities; Infrastructure and Health: Modeling Production, Processing and Distribution Infrastructure for a Resilient Food System. 2011. Urban Design Lab at the Earth Institute, Columbia University; Resilient Urban Food Systems: Opportunities, Challenges, and Solutions. 2013. Resilient Cities Team, ICLEI-Local Governments for Sustainability.
48. J. Kelly. nd. Building Local Food Systems to Increase Resiliency. Institute for Sustainable Communities. sustainablecommunitiesleadershipacademy

.org/resource_files/documents/building-local-food-systems-increase-resiliency.pdf

49. K. Clancy. 2013. High-priority research approaches for transforming U.S. food systems. *Journal of Agriculture, Food Systems, and Community Development*, 3(4), 5–7.
50. D. Liverman and J. Ingram. 2010. Why Regions? In Ingram et al. (eds). *Food Security and Global Environmental Change.* 2010. Earthscan: Washington, DC.
51. A. Tagtow and S. Roberts. 2011. Cultivating Resilience: An Iowa Food System Blueprint that Advances the Health of Iowans, Farms and Communities. Iowa Food Systems Council. IowaFoodSystemsCouncil.org/cultivating-resilience/
52. The Iowa Food Systems Council has completed three projects since the release of the 2011 plan: a marketing campaign to encourage gardening by low-resource individuals and gardeners to grow and donate extra produce to food pantries and other community-based organizations, and plans for value-added processing of fruits, vegetables and poultry. CultivateIowa.org.
53. E. Campbell. 2011. Regional Food System Plan for Vermont's Northeast Kingdom. Northeastern Vermont Development Association, St. Johnsbury, VT.
54. Vermont's Farm to Plate Investment Program is one of only a few statewide programs in the United States that specifically support the development of local, sustainable food systems. vtfoodatlas.com/plan/
55. E. Campbell. 2011. Figure 1.1, p. 4 in *Regional Food System Plan for Vermont's Northeast Kingdom*. Northeastern Vermont Development Association, St. Johnsbury, VT.
56. R. Carden. 2014. Feces Can Change the World: Using Human Excreta as Fertilizer. Food Tank. foodtank.com/news/2014/03/feces-can-change-the-world-using-human-excreta-as-fertilizer
57. Urine Nutrient Reclamation Project. Rich Earth Institute, Brattleboro, VT. richearthinstitute.org/
58. New England Food Policy: Building a Sustainable Food System. 2014. American Farmland Trust, Conservation Law Foundation, Northeast Sustainable Agriculture Working Group. newenglandfoodpolicy.org
59. Resilience theory proposes that as disturbance offers opportunities for innovation or transformation, these opportunities can be managed to bring about more desirable systems; see B. Walker, C. S. Holling, S. R. Carpenter and A. Kinzig. 2004. Resilience, adaptability and transformability in social-ecological systems, *Ecology and Society*, 9(2).
60. See chapter 4.
61. See chapter 9.

62. M. Miller et al. 2013. Critical Research Needs for successful food systems adaptation to climate change. *Journal of Agriculture, Food Systems and Community Development*. 3(4), 161–175.
63. Ibid.
64. Urban Design Lab. 2011. Infrastructure Health: Modelling Production, Processing and Distribution for a Resilient Regional Food System. Earth Institute, Columbia University.
65. R. Hoppe and D. Banker. 2010. Structure and Finances of US Farms: Family Farm Report, 2010 Edition. USDA-ERS Economic Information Bulletin No. 66.
66. Annual Estimates of the Resident Population: April 1, 2010 to July 1, 2013, U.S. Census Bureau, Population Division, Release Date: March 2014.
67. B. Walker and D. Salt. 2012. Managing Resilience. Ch. 4 in *Resilience Practice: Building Capacity to Absorb Disturbance and Maintain Function*. Island Press: Washington, DC.
68. Walthall et al. Adapting to Climate Change. Ch. 6 in *Climate Change and Agriculture in the United States: Effects and Adaptation*. 2012. USDA Technical Report 1935.
69. Ibid.
70. P. Ericksen et al. 2010. Adapting food systems. Ch. 8 in *Food Security and Global Environmental Change*. Earthscan: Washington, DC.
71. A. Iles and R. Marsh. 2012. Nurturing diversified farming systems in industrialized countries: How public policy can contribute. *Ecology and Society*, 17(4): 42.

Index

About the Author

Amy Moore

LAURA LENGNICK has explored agricultural and food system sustainability through more than 30 years of work as a federal researcher and policy-maker, college educator, community activist and farmer to understand what it takes to move sustainability values into action on the farm, in our communities and as a nation. Her research in sustainable agriculture systems was nationally recognized with a USDA Secretary's Honor Award in 2000 and she has broad federal policy expertise gained through work as a congressional lobbyist, a private consultant, and a US Senate staffer. For more than a decade, Laura led the academic program in sustainable agriculture at Warren Wilson College, where she also served as the Director of Sustainability Education, coordinated energy descent action planning, and developed an innovative sustainable dining policy for the college. She contributed to the third National Climate assessment as a lead author of the USDA report *Climate Change and Agriculture in the United States: Effects and Adaptation*. In 2014, Laura left Warren Wilson to serve as Co-Director of Resilience Initiatives for Second Nature. Laura is also an affiliated researcher with the Local Food Research Center and a climate resilience planning consultant with Fernleaf Solutions, both located in Asheville, NC.

If you have enjoyed *Resilient Agriculture*, you might also enjoy other

Books to Build a New Society

Our books provide positive solutions for people who want to make a difference. We specialize in:

Climate Change ◆ Conscious Community
Conservation & Ecology ◆ Cultural Critique
Education & Parenting ◆ Energy ◆ Food & Gardening
Health & Wellness ◆ Modern Homesteading & Farming
New Economies ◆ Progressive Leadership ◆ Resilience
Social Responsibility ◆ Sustainable Building & Design

New Society Publishers

ENVIRONMENTAL BENEFITS STATEMENT

New Society Publishers has chosen to produce this book on recycled paper made with 100% post consumer waste, processed chlorine free, and old growth free.

For every 5,000 books printed, New Society saves the following resources:[1]

38	Trees
3,413	Pounds of Solid Waste
3,755	Gallons of Water
4,898	Kilowatt Hours of Electricity
6,205	Pounds of Greenhouse Gases
27	Pounds of HAPs, VOCs, and AOX Combined
9	Cubic Yards of Landfill Space

[1]Environmental benefits are calculated based on research done by the Environmental Defense Fund and other members of the Paper Task Force who study the environmental impacts of the paper industry.

For a full list of NSP's titles, please call 1-800-567-6772 *or check out our web site at:*

www.newsociety.com